Encyclopedia of Soybean: Global Case Studies Volume II

Edited by **Albert Marinelli and Kiara Woods**

New York

Published by Callisto Reference,
106 Park Avenue, Suite 200,
New York, NY 10016, USA
www.callistoreference.com

Encyclopedia of Soybean: Global Case Studies
Volume II
Edited by Albert Marinelli and Kiara Woods

International Standard Book Number: 978-1-63239-297-8 (Hardback)

Contents

Preface

I am honored to present to you this unique book which encompasses the most up-to-date data in the field. I was extremely pleased to get this opportunity of editing the work of experts from across the globe. I have also written papers in this field and researched the various aspects revolving around the progress of the discipline. I have tried to unify my knowledge along with that of stalwarts from every corner of the world, to produce a text which not only benefits the readers but also facilitates the growth of the field.

Soybean is the most crucial oilseed and livestock feed crop in the world. Traversing the 10-year period from 2001 to 2010, world soybean production increased from 168 to 258 million metric tons (54% increase). The crop's high protein content (nearly 40% of seed weight) and oil content (approximately 20%) are characteristics that cannot be competed with any other agronomic crop. In lieu of soybean's extraordinary ascendancy, there has been an increased research interest in the crop throughout the world. Information in this book presents an inclusive outlook of research attempts in soybean genetics that will aid soybean stakeholders and scientists across the world.

Finally, I would like to thank all the contributing authors for their valuable time and contributions. This book would not have been possible without their efforts. I would also like to thank my friends and family for their constant support.

Editor

Soybean Genetics

In vitro Regeneration and Genetic Transformation of Soybean: Current Status and Future Prospects

Thankaraj Salammal Mariashibu,
Vasudevan Ramesh Anbazhagan, Shu-Ye Jiang,
Andy Ganapathi and Srinivasan Ramachandran

Additional information is available at the end of the chapter

1. Introduction

Soybean [*Glycine max* (L.) Merrill], grown for its edible seed protein and oil, is often called the miracle crop because of its many uses. It belongs to the genus Glycine under the family Leguminosae, and is widely cultivated in the tropics, subtropics and temperate zones of the world [1].

Soybean is now an essential and dominant source of protein and oil with numerous uses in feed, food and industrial applications. It is the world's primary source of vegetable oil and protein feed supplement for livestock. The global production of soybeans is 250-260 million tons per year. The US is the largest producer with 90.6 million metric tons. Other major countries such as Brazil, Argentina, China and India contributing 70, 49.5, 15.2 and 9.6 million metric tons, respectively [2]. The US, Brazil and Argentina are the major exporters of beans; while China and Europe are the major importers. The annual world market value is around 2 billion US dollars, which stands second in world food production.

Recent nutritional studies claim that consumption of soybean reduces cancer, blood serum cholesterol, osteoporosis and heart diseases [3]. This has sparked increased demand for the many edible soybean products. The priority for more meat in diets among the world's population has also increased the demand for soybean protein for livestock and poultry feed.

Soybean seeds are comprised of 40% protein, mostly consisting of the globulins β-conglycinin (7S globulin) and glycinin (11S globulin). The oil portion of the seed is composed primarily of five fatty acids. Palmitic and stearic acids are saturated fatty acids and comprise 15% of the oil. Soybean is rich in the unsaturated fatty acids like oleic, linoleic and linolenic, which make up 85% of the oil. Soybeans are a good source of minerals, B vitamins, folic acid and isoflavones, which are credited with slowing cancer development, heart diseases and osteoporosis [4].

The productivity of soybean has been limited due to their susceptibility to pathogens and pests, sensitivity to environmental stresses, poor pollination and low harvest index. Among the abiotic stresses, drought is considered the most devastating, commonly reducing soybean yield by approximately 40% and affecting all stages of plant growth and development; from germination to flowering, and seed filling and development as well as seed quality [5]. It suffers from many kinds of fungal diseases, such as frogeye leaf spot and brown spot [6]. As demand increases for soybean oil and protein, the improvement of soybean quality and production through genetic transformation and functional genomics becomes an important issue throughout the world [7].

The main objectives of soybean improvement include increase in yield, development of resistance to various insects, diseases and nutritional quality. Commercial breeding is still very important for the genetic improvement of the crop. However, breeding is difficult due to the fact that the soybean is a self pollinating crop, and the genetic base of modern soybean cultivars is quite narrow [8]. Most of the current soybean genotypes have been derived from common ancestors; therefore, conventional breeding strategies are limited in capability to expand the soybean genetic base. Recent advances in *in vitro* culture and gene technologies have provided unique opportunities for the improvement of plants, which are otherwise difficult through conventional breeding. The technology of plant transformation is only moderately or marginally successful in many important cultivars of crops, which can be a major limiting factor for the biotechnological exploitation of economically important plant species and the wider application of genomics.

Although numerous methods have been developed for introducing genes into plant genomes, the transformation efficiency for soybean still remains low [9]. Since the first successful transformation of soybean was reported [10], two major methods have been used in soybean transformation: one is particle bombardment of embryogenic tissue and another is *Agrobacterium tumefaciens*-mediated transformation of the cotyledonary node. Both methods have limitations: the former is highly genotype-dependent, requires a prolonged tissue culture period and tends to produce multiple insertion events, while the latter is labour intensive and requires specially trained personnel to undertake the work [9]

For soybean *in vitro* regeneration, two principal methods have been identified: somatic embryogenesis and shoot morphogenesis. Each of these systems presents both advantages and disadvantages for production of transformed plants, and each can be used with both of the predominant transformation systems [11]. A better understanding of physiology and molecular biology of *in vitro* morphogenesis needs focal attention to reveal their recalcitrant nature.

The present review gives an overview on the problems associated with low transformation efficiency, and the research conducted to improve tissue culture and transformation efficiency of soybean during the past (Table 1&2) and also discuss the future prospects, demands of these technologies and upcoming new technologies in soybean improvement.

Year	Explant tissue	Major contribution	Reference
1973	Hypocotyl	Adventitious bud development	Kimball and Bingham, [13]
1980	Cotyledonary node	Shoot morphogenesis	Cheng et al. [14]
1986	Immature embryo	Plant regeneration from callus	Barwale et al. [18]
1986	Cotyledonary node	Multiple shoot formation	Barwale et al. [19]
1986	Cotyledonary node	Multiple shoot formation	Wright et al. [20]
1987	Epicotyl	Callus induction and shoot regeneration	Wright et al. [29]
1988	Cotyledonary node	Transfered npt II and gus gene by Agrobacterium mediated transformation	Hinchee et al. [10]
1988	Immature seeds	Developed transgenic soybean by Particle bombardment	McCabe et al. [25]
1989	Germinating seeds	Transfered npt II gene by Agrobacterium mediated transformation	Chee et al. [45]
1989	Immature seed	Particle bombardment of meristems	Christou et al. [62]
1990	Immature cotyledon	Plant regeneration from protoplast	Luo et al. [127]
1990	Cotyledon, cotyledonary node	Evaluated Agrobacterium sensitivity and adventitious shoot formation	Delzer et al. [44]
1990	Immature cotyledon, plumule, cotyledonary node	Analysed plant regeneration efficiency of various explants	Yang et al. [32]
1990	Immature embryo	Organogenesis and plant regeneration	Yeh,[128]
1990	Primary leaf node	Adventitious shoot formation	Kim et al. [27]
1991	Immature cotyledon	Plant regeneration from protoplast	Dhir et al. [129]
1992	Epicotyl and hypocotyl	Investigated the stimulative effect of allantoin and amides on shoot regeneration	Shetty, et al. [21]
1993	Shoot tip	Transfered gus gene via particle bombardment	Sato et al. [130]
1994	Primary leaf node	Investigated the synergistic effect of proline and micronutrients on shoot regeneration	Kim et al. [40]
1996	Cotyledonary node	Developed transgenic soybean resistance to bean pod mottle virus (BPMV)	Di et al. [131]
1997	Cotyledonary node and hypocotyl	Multiple shoot induction by TDZ	Kaneda et al. [22]
1998	Cotyledonary node	Evaluation of sonication assisted Agrobacterium mediated transformation (SAAT) for cotyledonary node	Meurer et al. [50]
1998	Hypocotyl	Adventitious shoot regeneration	Dan and Reichert, [33]

Year	Explant tissue	Major contribution	Reference
1999	Cotyledonary node	Assessed the use of glufosinate as a selective agent in *Agrobacterium*-mediated transformation of soybean	Zhang et al. [61]
2000	Cotyledonary node	*Agrobacterium* two T-DNA binary system as a strategy to derive marker free transgenic soybean	Xing et al. [132]
2000	Cotyledonary node	Evaluated the effect of glyphosate as a selective agent for *Agrobacterium* mediated cotyledonary node transformation system	Clemente et al. [60]
2000	Embryonic axes	Used of Imazapyr as selection agent for selection of meristematic soybean cells	Aragao et al. [47]
2001	Cotyledonary node	Investigated the use of thiol compound to increase transformation frequency	Olhoft et al. [56]
2001	Cotyledonary node	Increased *Agrobacterium* infection using L-cystine	Olhoft and Somers, [16]
2001	Cotyledonary node	Developed transgenic soybean plants resistant to soybean mosaic virus (SMV)	Wang et al. [133]
2001	Cotyledonary node	Expressed oxalate oxidase gene for resistant to sclerotinia stem rot caused by *Sclerotinia sclerotiorum*	Donaldson et al. [65]
2003	Hypocotyl	Screened soybean genotype for adventitious organogenic regeneration	Reichert et al. [41]
2003	Cotyledonary node	Assessed the effect of genotype, plant growth regulators and sugars on regeneration from calli	Sairam et al. [1]
2003	Cotyledonary node	Used mixture of thiol compounds and hygromycin based selection for increased transformation efficiency	Olhoft et al. [57]
2004	Cotyledonary node	Assessed glufosinate selection for increased transformation efficiency	Zeng et al. [134]
2004	Cotyledonary node	Investigated the effect of seed vigor of explant source, selection agent and antioxidant on *Agrobacterium* mediated transformation efficiency	Paz et al. [15]
2004	Cotyledonary node	Transferred chitinase gene and the barley ribosome-inactivating protein gene to enhance fungal resistance	Li et al. [6]
2004	Mature and immature cotyledon	Shoot regeneration	Franklin et al. [31]
2004	Embryonic tip	Established regeneration and *Agrobacterium* mediated transformation system	Liu et al. [35]

Year	Explant tissue	Major contribution	Reference
2004	Cotyledonary node	Established liquid medium based system for selection transformed plants	Yun, [58]
2005	Cotyledonary node	Developed repetitive organogenesis system	Shan et al. [23]
2005	Cotyledonary node	Expressed Escherichia coli K99 fimbriae subunit antigen in soybean to use as edible vaccine	Piller et al. [66]
2006	Cotyledonary node	Agrobacterium mediated transformation efficiency was improved by using half seed explant from mature seed	Paz et al. [24]
2007	Cotyledonary node	Investigated Agrobacterium rhizogen to transform soybean cotyledonary node cells.	Olhoft et al. [59]
2007	Cotyledonary node	Expressed synthetic Bacillus thuringiensis cry1A gene that confers a high degree of resistance to Lepidopteran Pests	Miklos et al. [135]
2007	Cotyledonary node and leaf node	Established organogenic callus induction and Agrobacterium mediated transformation	Hong et al ., [43]
2007	Half seed	Expressed jasmonic acid carboxyl methyltransferase in soybean to produce methyl jasmonate, which resulted in tolerant to water stress	Xue et al. [67]
2008	Hypocotyl	Used silver nitrate to enhance adventitious shoot regeneration after Agrobacterium transformation and developed transgenic soybean producing high oleic acid content by silencing endogenous GmFAD2-1 gene by RNAi	Wang and Xu, [7]
2008	Cotyledonary node	Improved transformation efficiency using surfactant Silwet L-77 during Agrobacterium infection and L-cysteine during co-cultivation	Liu et al. [136]
2008	Cotyledonary node	Developed rapid regeneration system using whole cotyledonary node	Ma and Wu, [2008]
2010	Cotyledonary node	Production of isoflavone in callus cell lines by expression of isoflavone synthase gene.	Jiang et al. [69]
2010	Cotyledon and embryo	Developed shoot regeneration from calli of soybean cv.Pyramid	Joyner et al. [39]
2011	Hypocotyl	Transgenic soybean with low phytate content	Yang et al. [70]
2011	Cotyledon	Developed transgenic soybean with increased Vitamin E content by transferring γ-tocopherol methyltransferase (γ-TMT) gene in to seedling cotyledon	Lee et al. [137]

Table 1. Major landmarks in soybean organogenesis and transformation

Explant Tissue	Year	Major Contribution	Reference
Embryonic axes	1983	Embryoids development and plant regeneration *via* suspension culture	Christianson et al.[77]
Immature cotyledon	1984	Somatic embryo Induction	Lippmann & Lippmann, [84]
Immature cotyledon	1985	Plant regeneration *via* somatic embryogenesis	Lazzeri et al. [138]
Immature embryo	1985	Somatic embryogenesis and assessment of genotypic variation	Ranch et al. [139]
Immature embryo, cotyledon and, hypocotyl from germinating seedling	1986	Somatic embryogenesis from callus	Ghazi et al. [140]
Hypocotyl and cotyledon	1986	Embryoids development in suspension culture	Kerns et al. [141]
Immature embryo and cotyledon	1987	Investigated the effect of nutritional, physical, and chemical factors on somatic embryogenesis	Lazzeri et al. [85]
Immature cotyledon	1988	Investigated the effect of auxin and orientation of explant on somatic embryogenesis	Hartweck et al. [142]
Immature cotyledon	1988	Analysed genotype dependency and High concentration of auxin on somatic embryo induction	Komatsuda and Ohyama, [143]
Immature cotyledon	1988	Investigated the interaction between auxin and sucrose during somatic embryogenesis	Lazzeri et al. [86]
Immature cotyledon	1988	Germination frequency of somatic embryo has been improved by reducing the exposure to auxin	Parrott et al. [87]
Immature cotyledon	1988	Developed rapid growing maintainable embryogenic suspension culture	Finer and Nagasawa, [82]
Immature cotyledon	1988	Histological analysis to investigate secondary somatic embryo formation.	Finer, [79]
Immature cotyledon	1989	Demonstrated the effect of genotype on embryogenesis	Parrott et al. [144]
Immature cotyledon	1989	Developed primary transformants expressing zein gene by agrobacterium mediated transformation	Parrott et al. [105]
Immature cotyledon	1989	Assayed somatic embryo maturation for conversion into plantlets	Buchheim et al. [94]
Immature cotyledon	1989	Investigated the developmental aspects of somatic embryogenesis	Christou and Yang, [145]
Immature cotyledon	1990	Screened soybean genotypes for somatic embryo production	Komatsuda et al. [146]
Immature cotyledon	1991	Transformed embryogenic cultures with *gus* and *hpt* gene *via* particle bombardment	Finer and McMullen., [64]

Explant Tissue	Year	Major Contribution	Reference
Immature cotyledon	1991	Analysed the interaction between genotype and sucrose concentration on somatic embryogenesis	Komatsuda et al. [147]
Immature cotyledon	1991	Demonstrated adventitious shoot formation from cotyledonary and torpedo stage embryo	Wright et al. [148]
Immature cotyledon	1992	Somatic embryo proliferation by somatic embryo cycling.	Liu et al. [83]
Immature cotyledon	1993	Improved germination efficiency of somatic embryos of cultivar H7190 by desiccation	Bailey et al. [101]
Immature cotyledon	1993	Demonstrated genotypic effect on induction, proliferation, maturation and germination of somatic embryo	Bailey et al. [96]
Immature cotyledon	1993	Investigated the factors affecting somatic embryogenesis	Lippmann & Lippmann, [149]
Immature cotyledon	1993	Soybean transformation by particle bombardment of embryogenic cultures	Sato et al. [130]
Immature cotyledon	1994	Developed transgenic soybean resistance to insect.	Parrott et al. [150]
Immature embryos	1995	Investigated the effect of glutamine and sucrose on dry matter accumulation and composition of somatic embryo.	Saravitz and Raper, [151]
Immature cotyledon	1996	Demonstrated the significance of embryo cycling for transformation	Liu et al.[152]
Immature cotyledon	1996	Transformed embryogenic cultures with 12 different plansmid via particle bombardment	Hadi et al. [115]
Immature cotyledon	1996	Developed transgenic soybean expressing a synthetic Bacillus thuringiensis insecticidal crystal protein gene (BtcrylAc) which is resistance to insects	Stewart et al. [46]
Immature cotyledon	1997	Investigated the effect of ethylene inhibitors on embryo histodifferentiation and maturation	Santos et al. [92]
Epicotyls and primary leaves	1997	Somatic embryogenesis and plant regeneration from cotyledon, epicotyls and primary leaves	Rajasekaran and Pello, [153]
Immature cotyledon	1997	Studied the effect of explant orientiation, pH, solidifying agent and wounding on induction of soybean from immature cotyledons	Santarém et al. [81]
Immature cotyledon	1998	Studied growth characteristics of embryogenic cultures for transformability	Hazal et al. [113]

Explant Tissue	Year	Major Contribution	Reference
Immature cotyledon	1998	Established sonication-assisted *Agrobacterium* mediated transformation of soybean immature cotyledon	Santarem et al.[48]
Immature cotyledon	1998	Established sonication-assisted Agrobacterium mediated transformation of embryogenic suspension culture tissue	Trick and Finer, [108]
Immature cotyledon	1998	Improved proliferation efficiency of embryogenic cultures by modifying sucrose and nitrogen content in medium	Samoylov et al. [89]
Immature cotyledon	1998	Developed liquid medium based system for histodifferentiation of embryogenic cultures	Samoylov et al. [154]
Immature cotyledon	1998	Studied soluble carbohydrate content in soybean somatic and zygotic embryo during development.	Chanprame et al. [155]
Immature cotyledon	1999	Studied the factors influencing transformation of prolific embryogenic cultures using bombardment	Santarem and Finer, [116]
Immature cotyledons	1999	Developed transgenic plants with bovine milk protein, β-casein	Maughan et al. [114]
Immature cotyledons	1999	Transformed GFP into embryogenic suspension culture with the aim to improve transformation and regeneration strategy	Ponappa et al. [156]
Immature cotyledons	2000	Improved somatic embryo development and maturation by application of ABA	Tian and Brown, [157]
Immature cotyledon	2000	Screened genotypes for proliferative embryogenesis	Simmonds and Donaldson, [97]
Immature cotyledons	2000	Studied physical factors influencing somatic embryo development from immature cotyledons.	Bonacin et al. [99]
Immature cotyledon	2000	Investigated the factors affecting *Agrobacterium* mediated transformation soybean	Yan et al. [109]
Immature cotyledon	1989	Investigated maturation of somatic embryo for efficient conversion into plantlets	Buchheim et al. [94]
Immature cotyledon	2000	Developed and evaluated transgenic soybean expressing a synthetic cry1Ac gene from *Bacillus thuringiensis* for resistance to variety of insects	Walker et al. [158]
Immature cotyledon	2001	Effect of polyethylene glycol and sugar alcohols on soybean somatic embryo germination and conversion	Walker and Parrott, [90]

Explant Tissue	Year	Major Contribution	Reference
Immature cotyledon	2000	Developed integrated bombardment and Agrobacterium transformation method	Droste et al.[159]
Immature cotyledon	2001	Screened soybean from different location in the US for uniform embryogenic response	Meurer et al. [103]
Immature cotyledon	2001	Studied the effect of osmotica for their influence on embryo maturation and germination	Walker & Parrott, [90]
Immature cotyledon	2001	Developed transgenic plant expressing 15-kD zein protein under β-phaseolin seed specific promoter	Dinkins et al. [125]
Immature cotyledon	2001	Somatic embryogenesis in Brazilian soybean cultivars	Droste et al. [160]
Immature cotyledon	2002	Somatic embryogenesis and particle bombardment for south Brazil cultivars	Droste et al. [100]
Immature cotyledon	2002	Histological analysis of developmental stages of somatic embryogenesis	Fernando et al. [161]
Immature cotyledon	2002	Screened soybean genotypes for somatic embryo induction and maturation capability	Tomlin, [162]
Immature cotyledon	2003	Investigated the effect of proliferation, maturation and desiccation on somatic embryo conversion	Moon and Hildebrand, [88]
Immature cotyledon	2004	Improved transformation efficiently using Agrobacterium strain KYRT1 carrying pKYRTI	Ko et al. [111]
Immature cotyledon	2004	Developed transgenic plant containing phytase gene that store (produces) more phosphrous in seed.	Chiera et al. [163]
Immature cotyledon	2004	Developed fertile transplastomic soybean	Dufourmantel et al.[117]
Immature cotyledon	2004	Transferred chi and rip gene to enhance fungal resistance	Li et al. [6]
Immature cotyledon	2004	Improved transformation efficiency using Agrobacterium strain KYRT1	Ko and Korban, [80]
Immature cotyledon	2004	Analysed media components and pH on somatic embryo induction	Hoffmann et al. [80]
Immature cotyledon	2005	Developed transgenic soybean expressing maize γ-zein protein	Li et al. [124]
Immature cotyledon	2005	Modified soybean histodifferentiation and msaturation medium with the aim to improve the protein and lipid composition of somatic embryo	Schmidt et al. [164]
Immature cotyledon	2005	Analysed the effect of carbon source and polyethylene glycol on embryo conversion	Korbes et al. [91]

Explant Tissue	Year	Major Contribution	Reference
Immature cotyledon	2006	Improved fatty acid content	Chen et al. [119]
Immature cotyledon	2006	Investigated the ontogeny of somatic embryogenesis	Santos et al. [165]
Somatic embryo	2006	Developed transgenic soybean resistance to dwarf virus	Tougou et al. [120]
Immature cotyledon	2006	Investigated the influence of antibiotics on embryogenic cultures and *Agrobacterium tumefaciens* suppression in soybean transformation	Wiebke et al. [166]
Immature cotyledon	2006	Developed transgenic soybean for increased production of ononitol and pinitol	Chiera et al. [167]
Immature cotyledon	2007	Developed transgenic soybean resistant to dwarf virus	Tougou et al. [168]
Immature cotyledon	2007	Improved somatic embryogenesis in recalcitrant cultivars by back cross with a highly regenerable cultivar Jack	Kita et al. [104]
Immature cotyledon	2007	Evaluated Japanese soybean genotypes for somatic embryogenesis	Hiraga et al. [102]
Immature cotyledon	2007	Soybean seed over expressing the *Perilla frutescens* γ-tocopherol methyltransferase gene	Tavva et al. [123]
Immature cotyledon	2007	Improved protein quality in transgenic soybean transformed with modified Gy1 proglycinin gene with a synthetic DNA encoding four continuous methionines.	El-Shemy et al. [169]
Immature cotyledon	2007	Analysed the effect of Abscisic acid on somatic embryo maturation and conversion.	Weber et al. [170]
Immature cotyledon	2007	Developed transgenic soybean resistance to soybean mosaic virus	Furutani et al. [121]
Immature cotyledon	2008	Used a new Selectable Marker Gene Conferring resistance to Dinitroanilines	Yemets et al. [171]
Immature cotyledon	2008	Developed strategy for transfer of multiple genes *via* micro projectile-mediated bombardment	Schmidt et al. [172]
Immature cotyledon	2009	Assessed the effect mannitol, abscisic acid and explant age on somatic embryogenesis in Chinese soybean cultivars	Yang et al. [98]
Somatic embryo	2009	Developed transgenic soybean with increased oil content	Rao and Hildebrand, [118]

Explant Tissue	Year	Major Contribution	Reference
Embryonic tip	2010	somatic embryogenesis and plant regeneration from the immature embryonic shoot tip	Loganathan et al. [173]
Immature cotyledon	2010	Developed transgenic soybean with more tryptophan content in seed	Ishimoto et al. [122]
Immature cotyledon	2010	Screening of Brazilian soybean genotypes for embryogenesis	Droste et al. [174]
Immature cotyledon	2011	Demonstrated Metabolic engineering of soybean seed coat for the production of novel biochemicals	Schnell et al. [126]
Immature cotyledon	2011	Investigated developmental profile of storage reserve accumulation in soybean somatic embryos	He et al. [175]
Immature cotyledon	2011	Improved transformation efficiency by Micro wounding with DNA free particle bombardment followed by Agrobacterium mediated transformation.	Wiebke et al. [112]
Immature cotyledon	2012	Developed vacuum infiltration assisted *Agrobacterium* mediated transformation for Indian soybean cultivars.	Mariashibu et al. [176]

Table 2. Major landmarks in soybean somatic embryogenesis and transformation

2. Organogenesis and transformation

Organogenesis is characterized by the production of a unipolar bud primordium with subsequent development of the primordium into a leafy vegetative shoot. A successful plant regeneration protocol requires appropriate choice of explant, definite media formulations, specific growth regulators, genotype, source of carbohydrate, gelling agent, other physical factors including light regime, temperature, humidity and other factors [12]. Plant regeneration by organogenesis in soybean was first reported by Kimball and Bingham, [13] from hypocotyl sections followed by Cheng et al.[14] by culturing seedling cotyledonary node segments. Transfer of T-DNA into cotyledonary node cells by *Agrobacterium* mediated transformation was first reported by Hinchee et al. [10]. Advancement in soybean transformation appears to be slow compared to some of the recent improvement in cereal transformation (Paz et al. 2004). Olhoft et al. [16] stated that the efficiency of soybean transformation has to be improved 5-10 times before one person can produce 300 transgenic lines per year. Soybean transformation efficiency has been improved by optimizing the selection system, enhancing explant-pathogen interaction and improving culture conditions to promote regeneration and recovery of transformed plants.

2.1. Organogenesis

The successful application of biotechnology in crop improvement is based on efficient plant regeneration protocol. Soybean has been considered as recalcitrant to regenerate *in vitro*. Tissue culture responses are greatly influenced by three main factors viz. whole plant physiology of donor, *in vitro* manipulation, and *in vitro* stress physiology [17]. After the first report of adventitious bud regeneration from hypocotyl sections by Kimball and Bingham, [13] researchers have used different parts of the soybean plant as explants for successful shoot morphogenesis in soybean. These include cotyledonary node [10,14,18-24], shoot meristems [25], stem-node [26,27] epicotyls [28], primary leaf [29], cotyledons [30,31], plumules (32), hypocotyls [22,33,34], and embryo axes [25,35]. Plant regeneration *via* organogenesis from cotyledonary node was found to be the most convenient and faster approach in soybean. However, much improvement is needed for the cotyledonary node regeneration system. This limitation is mainly due to low frequency of shoot regeneration, long regeneration period and explant growth difficulties, which prevent the plant from being regeneration-competent[36].

The nutritional requirement for optimal shoot bud induction from different explants has been reported to vary with mode of regeneration. Media compositions have a key role in shoot morphogenesis, the basal medium MS [37] is most commonly used for soybean organogenesis and the medium B5 [38] are useful in some approaches. Benzylaminopurine (BA) has been the most commonly used plant growth regulator either alone or in combination with a low concentration of cytokinins, kinetin or thidiazuron (TDZ) [22, 39]. TDZ was reported to induce multiple bud tissue (MBT) from cotyledonary node axillary meristem which then gives shoots in the presence of BA [23]. The efficiency of shoot bud formation were enhanced by supplementing media with proline, increased level of MS micro nutrients [40], and ureide in the form of allantoin and amides [21].

Adventitious shoot regeneration from cotyledonary node or leaf node is based on proliferation of meristems. Use of pre-existing shoot meristems in transformation procedures can increase the chance of chimerism, so identifying tissues that can produce shoots in the absence of such pre-formed organs would be important [41]. Adventitious soybean shoots have been induced from hypocotyls [13]; cotyledons [18, 20], primary leaves [29] and epicotyls [28]. Hypocotyls of seedlings have been used as explants for adventitious shoot regeneration by Kaneda et al. [22]. Explants cultured on media supplemented with TDZ induced adventitious shoots more efficiently than BA. Histological analysis of adventitious shoot regeneration from the hypocotyl shows shoot primordias, formed from parenchymatous tissues of central pith and plumular trace regions [33]. Hypocotyls of seedlings have seldom been used as explants, even though the shoot regeneration frequency from hypocotyl segments was found to be higher than from cotyledons [22]. Franklin et al. [31] investigated the factors affecting adventitious shoot regeneration from the proximal end of mature and immature cotyledons. The presence of BAP and TDZ in the medium exerted a synergistic effect, in that regeneration efficiency was higher than for either cytokinin alone.

Indirect organogenesis is important as an alternative source of genetic variation in order to recover somaclones with interesting agronomic traits. Callus regeneration is advantageous over direct regeneration for transformation since effective selection of transgenic cells can be achieved [1]. However, the efforts made to regenerate plants from callus have yielded poor results since plants could not be regenerated from any type of soybean callus [42]. Yang et al. [32] compared different explants excised from immature and germinated seeds for callus mediated organogenic regeneration, although induction of organogenic callus was easily achieved by culture of immature cotyledons, development of adventitious buds from these calluses and the subsequent growth of these buds to shoots were inefficient, suggesting that only part of the callus was competent for regeneration. Sairam et al. [1] developed a rapid and efficient protocol for regeneration of genotype-independent cotyledonary nodal callus for cultivars Williams 82, Loda and Newton through manipulation of plant growth regulators and carbohydrates in the medium. Hong et al. [43] reported organogenic callus induction from cotyledonary node and leaf node explants in media supplemented with TDZ and BA, the system has been successfully utilized for *Agrobacterium*-mediated transformation

2.2. Genotype

Among the different factors affecting soybean regeneration, the genotypic dependence is ranked quite high. Since there is strong genotype specificity for regeneration of different soybean genotypes, a major limiting factor, it is pivotal to formulate genotype specific regeneration protocols. Genotype specificity for regeneration in soybean is well documented, although organogenesis is less genotype dependent and has become routine in several laboratories [18,20,28,29&33]. Reichert et al. [41] tested organogenic adventitious regeneration from hypocotyl explants excised from 18 genotypes. Plant formation from hypocotyl explants showed that all genotypes were capable of producing elongated shoots that could be successfully rooted. This study confirmed the genotype independent nature of this organogenic regeneration from the hypocotyl explant. Sairam et al. [1] developed an efficient genotype independent cotyledonary nodal callus mediated regeneration protocol for soybean cultivars Williams 82, Loda and Newton developed through manipulation of plant growth regulators and carbon source. Callus induction and subsequent shoot bud differentiation were achieved from the proximal end of cotyledonary explants on modified MS [37] media containing 2,4-dichlorophenoxyacetic acid (2,4-D) and benzyladenine (BA), respectively. Sorbitol was found to be the best for callus induction and maltose for plant regeneration. The genotypic dependence of regeneration from cotyledon explants could be reduced by the use of combinations of cytokinins (Franklin et al. [31]). Though there was no significant difference in shoot bud formation among different genotypes, but there was significant difference in conversion of the number of regenerated plants in each cultivar (Delzer et al. [44]).

2.3. *Agrobacterium* mediated transformation

Agrobacterium-mediated transformation of soybean was first demonstrated by Hinchee et al. [10] through delivering, T-DNA into cells in the axillary meristems of the cotyledonary-node. After that scientists have attempted to introduce a lot of genes using *Agrobacterium* [25, 45-47]. The cotyledonary-node method is a frequently used soybean transformation system based on *Agrobacterium*-mediated T-DNA delivery into regenerable cells in the axillary meristems of the cotyledonary-node [16]. The efficiency of this transformation system remains low, apparently because of infrequent T-DNA delivery to cells in the cotyledonary-node axillary meristem, inefficient selection of transgenic cells that give rise to shoot meristems, and low rates of transgenic shoot regeneration and plant establishment. The development of an effective Agrobacterium transformation method for soybean depends on several factors including plant genotype, explant vigor, Agrobacterium strain, vector, selection system, and culture conditions [48, 49]. Increased soybean transformation efficiency, may be achieved by further optimizing the selection system, enhancing explant-pathogen interaction and improving culture conditions to promote regeneration and recovery of transformed plants. It has been reported that soybean genotype contributed to variation in susceptibility to *Agrobacterium* and regenerability in tissue culture [50, 51]. In addition, surface sterilization of plant tissue material for *in vitro* tissue culture and transformation is one of the critical steps in carrying out transformation experiments. While a short time of sterilization cannot completely decontaminate explants, prolonged sterilization may cause damage to explants and consequently affect their regenerability [52]. Antioxidant reagents such as cysteine, dithiothreitol, ascorbic acid and polyvinyl pyrrolidone have been used in plant transformation optimization to enhance either tissue culture response or transformation efficiency [53-55]. Recently, high transformation efficiency has also been reported in soybean by adding cysteine and thiol compounds to the cocultivation media [16, 56,57]. Liu et al. [35] established *Agrobacterium* mediated transformation using shoot tip explants of Chinese soybean cultivars. It had the advantage over the cotyledonary node by having no necrosis after infection, and showed more transient *gus* expression as embryonic tips are more sensitive to *Agrobacterium* because they contain promeristems and procambium. Yun, [58] established liquid medium to select transformed plants from the cotyledonary node. Liquid selection has proven to be more efficient than solid selection due to the direct contact of the explants with the medium and the selection agent in the medium. Olhoft et al. [59] transformed soybean cotyledonary nodes using *Agrobacterium rhizogens* strain SHA17 for the first time. The transformation efficiency was as high as 3.5 fold when compared with *Agrobacterium tumefaciens* strain AGL1. Clemente et al. [60] successfully used and evaluated the effect of glyphosate as a selective agent within the *Agrobacterium mediated* cotyledonary transformation system. Imazapyr is a herbicidal molecule that inhibits the enzymatic activity of acetohydroxyacid synthase, which catalyses the initial step in the biosynthesis of isoleucine, leucine and valine. Aragao et al. [47] used Imazapyr as a selection agent for selection of meristematic soybean cells transformed with the *ahas* gene from Arabidopsis. The *bar* gene encodes for phosphinothricin acetyltransferase (PAT) which detoxifies glufosinate, the active ingredient in the herbicide. Zhang et al. [61] successfully used glyphosate to select transformed cells after *Agrobacterium* transformation of cotyledonary node cells.

2.4. Particle bombardment

Even though particle bombardment is a widely used technique for transforming soybean embryogenic cultures, it was rarely explored for shoot morphogenesis. McCabe et al. [25] was the first to report particle bombardment mediated transformation in soybean. Transforming meristems of soybean bu DNA coated gold particles followed by shoot regeneration in the presence of cytokinin, resulting in the development of chimeras. In subsequent studies, non-chimeric plants were obtained through the use of screening methods for the selection of plants that contained transgenic germ-line cells [32,62&63]. Shoot apex transformation is labour intensive because the meristematic tissue is diffcult to target and, without selection, a large number of plants must be regenerated and analysed [64].

2.5. Genes for trait improvement

Soybean has been improved by *Agrobacterium* mediated transformation followed by shoot regeneration. Wheat germin gene (gf-2.8) encoding an oligomeric protein and oxalate oxidase (*oxo*) genes were introduced into soybean to improve resistance to the oxalate-secreting pathogen *Sclerotina sclerotiorum* [65]. Li et al.[6] successfully utilized *Agrobacterium*-mediated transformation to transfer chitinase gene (*chi*) and the barley ribosome-inactivating protein gene (*rip*) into soybean cotyledonary node cells. Piller et al. [66] investigated the feasibility of expressing the major Enterotoxigenic Escherichia coli K99 fimbrial subunit, FanC, in soybean for use as an edible subunit vaccine. Xue et al. [67] successfully expressed jasmonic acid carboxyl methyltransferase (NTR1) gene from *Brassica campestris* into soybean cv.Jungery that produces methyl jasmonate and showed tolerance to water stress. Soybean oil contains very low level of α-tocopherol which is the most active form of tocopherol. The tocopherols present in the seed are converted into α- and β-tocopherols by overexpressing γ-tocopherol methyltransferase from *Brassica napus* (BnTMT) [68]. Jiang et al. [69] transferred isoflavone synthase (IFS) gene into soybean callus using *Agrobacterium*-mediated transformation and the transgenic plants produced increased levels of the secondary metabolite, isoflavone. Transgenic soybean plant containing PhyA gene of *Aspergillus ficuum* exhibited a lower amount of phytate in different soybean tissues including the leaf, stem and root. This indicated that engineering crop plants with a higher expression level of heterologous phytase could improve the degradation of phytate and potentially in turn mobilize more inorganic phosphate from phytate and thus reduce phosphate load on agricultural ecosystems [70].

3. Somatic embryogenesis and transformation

Somatic embryogenesis is a process by which a plant somatic cell develops into a whole plant without gametic fusion but undergoes developmental changes as that of zygotic embryogenesis [71, 72]. The first demonstration of *in vitro* somatic embryogenesis was reported in *Daucus carota* by Reinert [73]. The concept of embryogenesis has drawn a lot of attention because of its significance in theory and practice. Primarily, somatic embryos can be produced easily and quickly, so that it provides an economical and easy way to study plant development. Secondly, synthetic seeds developed from somatic embryos open the possibility of developing high quality seeds and may allow us to produce seeds from those plants that require a long period for seed production. Somatic embryogenesis is also useful in plant genetic engineering since regeneration *via* somatic embryogenesis is frequently single of cell origin, resulting in a low response of chimeras and high a number of true transgenic regenerants [74, 75].

3.1. Somatic embryogenesis

The first record of soybean somatic embryogenesis was reported by Beversdorf & Bingham [76], followed by Christianson et al. [77] who regenerated plants through the method. The immature cotyledon is the preferred explant for soybean somatic embryogenesis as it has pre-determined embryogenic cells. Somatic embryogenesis is a multi-step regeneration process starting with the formation of proembryogenic cell mass, followed by somatic embryo induction, their maturation, desiccation and finally plant regeneration [78].

Soybean somatic embryos were induced from immature cotyledon explants cultured on medium containing high levels of 2,4-D [79]. Even though NAA induced somatic embryogenesis from immature cotyledons, the mean number of embryos produced on 2,4-D was significantly higher [80]. Explant orientation, pH, solidifying agent, and 2,4-D concentration have a synergic effect on somatic embryo induction [81]. The early-staged somatic embryos can be maintained and proliferated by subculturing the tissue on either semi-solid medium [79] or liquid suspension culture medium [82]. Somatic embryos incubated in a medium containing NAA do not proliferate so well as those produced on a medium containing 2,4-D [83]. Somatic embryos initiated on NAA are more advanced in embryo morphology than those induced on 2,4-D and the efficiency of somatic embryo induction was highest with a medium containing 2-3% sucrose. Cultures initiated on lower sucrose concentrations tended to produce a higher amount of friable embryos, while increased concentrations of this sugar impaired embryo induction [80,84-86]. Histodifferentiation and maturation of somatic embryos doesn't need exogenous auxin or cytokinins [87]. Indeed, poorly developed meristem or swollen hypocotyls may be an undesired outcome of the application of exogenous auxins and cytokinins, respectively. Moon and Hildebrand, [88] investigated the effects of proliferation, maturation, and desiccation methods on conversion of soybean somatic embryos to plants. Somatic embryos proliferated on solid medium showed a higher regeneration rate when compared with the embryos proliferated in liquid medium. The growth period of somatic embryo development can be reduced one month by culturing in a medium devoid of 2,4-D and B_5 vitamins. Carbon source is critical for embryo nutritional health and improves somatic embryo maturation. The effects of carbohydrates on embryo histodifferentiation and maturation on liquid medium were analyzed by Samoylov et al. [89]. FNL medium supplemented with 3% sucrose (FNL0S3) or 3% maltose (FNL0M3) were compared. Data indicated that sucrose promotes embryo growth and significantly increases the number of cotyledon-stage embryos recovered during histodifferentiation and maturation. However, the percentages of plants recovered from embryos differentiated and matured in FNL0S3 was lower than those grown in FNL0M3 (Samoylov et al. 1998b). The quality of somatic embryos can be positively influenced by a low osmotic potential in maturation medium [90, 91]. Carbohydrates can act as an osmotic agent. Polyethylene glycol 4000, mannitol and sorbitol were tested as supplements to a liquid Finer and Nagasawa medium-based histodifferentiation/maturation medium FNL0S3, for soybean (*Glycine max* L. Merrill) somatic embryos of 'Jack' and F138 or 'Fayette'[90]. Overall, 3% sorbitol was found to be the best of the osmotic supplements tested. The ability of histodifferentiation and conversion of somatic embryo have been improved by the use of ethylene inhibitor aminoethoxyvinylglycine [92].

The effects of ethylene on embryo histodifferentiation and conversion were genotype-specific. The germination frequency of soybean embryos is very low [93], and therefore, partial desiccation of somatic embryos was emphasised with a view to improving the germination frequency in soybean [87,94&95]. Desiccation induced a physiological state there by increase the germination ability of somatic embryos [87].

3.2. Genotype

Soybean somatic embryogenesis is highly genotypic when compared to organogenesis. The existence of strong genotype specificity in the regeneration capacity of the different cultivars represents a major limiting factor for the advancement of soybean biotechnology. The embryogenic efficiency of soybean was shown to be different among cultivars at each stage (induction, proliferation, maturation, germination) of somatic embryogenesis [92,96] and it is very challenging to identify genotypes highly responsive to all stages. Simmonds and Donaldson, [97] screened 18 short season soybean genotypes for proliferative embryogenesis. Five genotypes produced embryogenic cultures which were proliferative for at least 6 months. Yang et al. [98] screened 98 Chinese soybean varieties for somatic embryogenesis and selected 12 varieties based on their embryogenic capacity. The greatest average number of plantlets regenerated per explant (1.35) was observed in N25281. Bonacin et al. [99] demonstrated the influence of genotype on somatic embryogenic capability of five Brazilian cultivars. Droste et al. [100] reported somatic embryo induction, proliferation and transformation of commercially grown Brazilian soybean cultivars for the first time. Soybean somatic embryo conversion is genotype dependent; germination frequency of H7190 was approximately three fold lower than that of PI 417138 [101]. Hiraga et al. [102] examined the capacity for plant regeneration through somatic embryogenesis in Japanese soybean cultivars and identified Yuuzuru and Yumeyutaka as having high potential for somatic embryogenesis. Several cultivars were identified as uniformly embryogenic at the primary induction phase at all locations, among which Jack was the best [103]. Kita et al. [104] evaluated somatic embryogenesis, proliferation of embryogenic tissue, and regeneration of plantlets in backcrossed breeding lines derived from cultivar Jack and a breeding line, QF2. The backcrossed breeding lines exhibited an increased capacity for induction and proliferation of somatic embryos and were used successfully to generate transgenic plants.

3.3. *Agrobacterium* mediated transformation

Recovery of the first transgenic plant *via* somatic embryogenesis in soybean was reported by Parrott et al. [105]. Immature cotyledon tissues were inoculated with *Agrobacterium* strain which contained 15 kD zein gene and the neomycin phosphotransferase gene. The explants were placed on medium containing high auxin for somatic embryo induction. Three transgenic plants containing the introduced 15 kD zein gene were regenerated. Unfortunately, these plants were chimeric and the 15 kD zein gene was not transmitted to the progeny. Sonication-assisted *Agrobacterium*-mediated transformation (SAAT) of immature cotyledons tremendously improves the efficiency of *Agrobacterium* infection by introducing large numbers of micro wounds into the target plant tissue [48]. The highest GUS expression was obtained when immature cotyledons were sonicated for 2s in the presence of *Agrobacterium* followed by co-cultivation for 3 days. Trick and Finer, [108] successfully employed Sonication-assisted *Agrobacterium*-mediated transformation of embryogenic suspension culture tissue and when SAAT was not used, no transgenic clones were obtained. Yan et al. [109] demonstrated the feasibility of *Agrobacterium* mediated transformation of cotyledon tissue for the production of fertile transgenic plants by optimising the *Agrobacterium* concentration, using co-cultivation time and selecting proper explant. Ko and Korban, [110] investigated optimal conditions for induction of transgenic embryos followed by *Agrobacterium* mediated transformation. Using cotyledon explants from immature embryos of 5-8mm length, a 1:1 (v/v) concentration of bacterial suspension and 4-day co-cultivation period significantly increased the frequency of transgenic somatic embryos. The *Agrobacterium* tumefaciens strain KYRT1 harboring the virulence helper plasmid pKYRT1 induces transgenic somatic embryos at a high frequency from infected immature soybean cotyledons [111]. Recently, the successful recovery of a high number of soybean transgenic fertile plants was obtained from the combination of DNA- free particle bombardment and *Agrobacterium*-mediated transformation using proliferating soybean somatic embryos as targets [112].

3.4. Particle bombardment

Particle bombardment is a widely used technique for transformation of embryogenic cultures of soybean; the major advantage of this technique over *Agrobacterium* is the removal of biological incompatibilities. Particle bombardment in soybean was first reported by Finer and McMullen [64], in which embryogenic suspension culture tissue of soybean was bombarded with particles coated with plasmid DNAs encoding hygromycin resistance and β-glucuronidase. Analysis of DNA from progeny plants showed genetic linkage for multiple copies of introduced DNA. Using particle bombardment, fertile plants could be routinely produced from the proliferating transgenic embryogenic clones. Hazal et al. [113] studied growth characteristics and transformability of embryogenic cultures and found that cultures bombarded between 2-6 days after transfer to fresh medium showed more transient expression of the reporter gene. Histological analysis showed that the most transformable cultures had cytoplasmic-rich cells in the outermost layers of the tissue. Maughan et al. [114] bombarded embryogenic cultures with plasmid containing 630-bp DNA fragment encoding a bovine milk protein, β-casein. Hadi et al. [115] co-transformed 12 different plasmids into embryogenic suspension culture by particle bombardment. Hybridization analysis of hygromycin resistance clones verified the presence of introduced plasmid DNAs. Santarem and Finer [116] investigated the effect of desiccation of target tissue, period of subculture prior to bombardment and number of bombardments per target tissue for enhancement of transient expression of the reporter gene. Desiccation of proliferating tissue for 10 min, subculture on the same day prior to bombardment and three times bombardment on a single day enhanced the transient expression of β-glucuronidase [116]. Dufourmantel et al. [117] successfully transformed chloroplasts from embryogenic tissue of soybean using DNA carrying spectinomycin resistance gene (*aadA*) by bombardment. All transplastomic T0 plants were fertile and T1 progeny was uniformly spectinomycin resistant, showing the stability of the plastid transgene. *Droste* et al. [100] successfully transformed embryogenic cultures of soybean cultivars recommended for commercial growing in South Brazil by bombardment, and this opened the field for the improvement of this crop in this country by genetic engineering.

3.5. Genes for trait improvement

Li et al. [6] attempted to transform two antifungal protein genes (*chitinase* and ribosome-in-activating protein) by co-transformation. Transgenic soybeans expressing the Yeast SLC1 Gene showed higher oil content [118]. They reported that, compared to controls, the average increase in triglyceride values went up by 1.5% in transgenic somatic embryos and also found that a maximum of 3.2% increase in seed oil content was observed in a T3 line. Transfer of $\Delta6$ desaturase, fatty acid elongase and D5 desaturase into soybean under seed specific expression produced arachidonic acid (ARA) in seeds of soybean [119]. In an attempt to enhance soybean resistance to viral diseases, several groups successfully generated transgenic plants by expressing an inverted repeat of soybean dwarf virus SbDV coat protein (*CP*) genes [120], or soybean mosaic virus (SMV) coat protein gene [121]. The nutritional quality of soybean has been improved for enhanced amino acid, proteins and vitamin production by transgenic technology [114, 122, 123, 124, and 125]. The feasibility of genetically engineering soybean seed coats to divert metabolism towards the production of novel biochemicals was tested by transferring the genes phbA, phbB, phbC from *Ralstonia eutropha*. Each gene was under the control of the seed coat peroxidase gene promoter [126]. The analysis of seed coats demonstrated that polyhydroxybutyrate (PHB) was produced at an averge of 0.12% seed coat dry weight.

4. Conclusion and future prospects

As demands increase for soybean oil and protein, the improvement of soybean quality and production through genetic transformation and functional genomics becomes an important issue throughout the world. Modern genetic analysis and improvement of soybean heavily depend on an efficient regeneration and transformation process, especially commercially important genotypes. The transformation techniques developed until now till date do not allow high-throughput analyses in soybean functional genomics; though significant improvements have been made in the particle bombardment of embryogenic culture and *Agrobacetrium* mediated transformation of the cotyledonary node over the past three decades. However, routine recovery of transgenic soybean plants using either of these two transformation systems has been restricted to a few genotypes with no reports of transformation on other locally available commercial genotypes. Therefore, development of an efficient and consistent transformation protocol for other locally available commercial genotypes, will greatly aid soybean functional genomics and transgenic technology.

Author details

Thankaraj Salammal Mariashibu[1], Vasudevan Ramesh Anbazhagan[1], Shu-Ye Jiang[1], Andy Ganapathi[2] and Srinivasan Ramachandran[1*]

*Address all correspondence to: sri@tll.org.sg

1 Temasek Life Sciences Laboratory, 1 Research Link, the National University of Singapore, Singapore

2 Department of Biotechnology, Bharathidasan University, Tiruchirappalli, Tamilnadu, India

References

[1] Sairam RV, Franklin G, Hassel R, Smith B, Meeker K, Kashikar N, Parani M, Abed Dal, Ismail S, Berry K, Goldman SL. A study on effect of genotypes, plant growth regulators and sugars in promoting plant regeneration via organogenesis from soybean cotyledonary nodal callus. Plant Cell, Tissue and Organ Culture 2003, 75:79-85.

[2] http://www.soystats.com/2011/page_30.htm.

[3] Birt DF, Hendrich S, Anthony M and Alekel DL (2004) Soybeans and the prevention of chronic human disease. Soybeans: Improvement, Production and Uses. J. Specht and R. Boerma, Eds. 3rd ed. pp. 1047–1117 American Society of Agronomy, Madison, WI

[4] Wilson RF. Seed composition. In: H.R. Boerma and J.E. Specht (eds) Soybean: improvement, production and uses. 3rd ed. ASA, CSSA, SSA, Madison, WI 2004. 621-677.

[5] Manavalan LP, Guttikonda SK, Tran LS, Nguyen HT. Physiological and molecular approaches to improve drought resistance in soybean. Plant Cell Physiology 2009, 50:1260-1276.

[6] Li HY, Zhu YM, Chen Q, Conner RL, Ding XD, Li J, Zhang BB. Production of transgenic soybean plants with two anti-fungal protein genes via Agrobacterium and particle bombardment. 2004 48(3):367-374.

[7] Wang G, Xu Y. Hypocotyl-based *Agrobacterium*-mediated transformation of soybean (Glycine max) and application for RNA interference. Plant Cell Reports 2008, 27:1177-1184.

[8] Hiromoto DM, Vello NA. The genetic base of Brazilian soybean (Glycine max (L.) Merrill) cultivars. Brazil Journal of Genetics 1986, 2:295-306.

[9] Cao D, Hou W, Song S, Sun H, Wu C, Gao Y, Han T. Assessment of conditions affecting *Agrobacterium rhizogenes*-mediated transformation of soybean. Plant Cell Tissue and Organ Culture 2009, 96:45-52.

[10] Hinchee MA, Connor-Ward DV, Newell CA, McDonnell RE, Sato SJ, Gasser CS, Fischhoff DA, Re DB, Fraley RT, Horsch RB. Production of transgenic soybean plants using *Agrobacterium*-mediated DNA transfer. Nature Biotechnology 1988, 6:915-922.

[11] Trick HN, Dinkins RD, Santarem, ER, Di R, Samoylov V, Meurer CA, Walker DR, Parrott WA, Finer JJ, Collins GB. Recent advances in soybean transformation. Plant Tissue Culture and Biotechnology 1997, 3(1):9-24.

[12] Kothari SL, Joshi A, Kachhwaha S, Ochoa-Alejo N. Chilli peppers -A review on tissue culture and transgenesis. Biotechnology Advances 2010, 28:35-48.

[13] Kimball SL, Bingham ET. Adventitious bud development of soybean hypocotyl sections in culture. Crop Science 1973, 13:758-760.

[14] Cheng TY, Saka H, Voqui-Dinh TH. Plant regeneration from soybean cotyledonary node segments in culture. Plant Science Letters 1980, 19:91-99.

[15] Paz MM, Shou H, Guo Z, Zhang Z, Banerjee AK, Wang K. Assessment of conditions affecting Agrobacterium-mediated soybean transformation using the cotyledonary node explant. Euphytica 2004, 136:167-179.

[16] Olhoft PM, Somers DA. L.Cysteine increases *Agrobacterium*- mediated T-DNA delivery into soybean cotyledonary-node cells. Plant Cell Reports 2001, 20:706-711.

[17] Benson EE. Special symposium: *in vitro* plant recalcitrance, do free radicals have a role in plant tissue culture recalcitrance? In Vitro Cellular and Developmental Biology Plant 2000, 86:163-70.

[18] Barwale UB, Keans HR, Widholm JM. Plant regeneration from callus cultures of several soybean genotypes via embryogenesis and organogenesis. Planta 1986, 167:473-481.

[19] Barwale UB, Mayer Jr MM, Widholm JM. Screening of Glycine max and Glycine soja genotypes for multiple shoot formation at the cotyledonary node. Theoretical and Applied Genetics 1986, 72:423-428.

[20] Wright MS, Koehler SM, Hinchee MA, Carnes MG. Plant regeneration by organogenesis in Glycine max. Plant Cell Reports 1986, 5:150-154.

[21] Shetty K, Asano Y, Oosawa K. Stimulation of in vitro shoot organogenesis in Glycine max (Merrill.) by allantoin and amides. Plant Science 1992, 81:245-251.

[22] Kaneda Y, Tabei Y, Nishimura S, Harada K, Akihama T, Kitamura K. Combination of thidiazuron and basal media with low salt concentrations increases the frequency of shoot organogenesis in soybeans [Glycine max (L.) Merr.]. Plant Cell Reports 1997, 17 8-12.

[23] Shan Z, Raemakers K, Tzitzikas EN, Ma Z, Visser RGF. Development of a highly effi-
      cient, repetitive system of organogenesis in soybean (Glycine max (L.) Merr). Plant
      Cell Reports 2005, 24:507-512.

[24] Paz MM, Martinez JC, Kalvig AB, Fonger TM, and Wang K. Improved cotyledonary
      node method using an alternative explant derived from mature seed for efficient
      Agrobacterium mediated soybean transformation. Plant Cell Reports 2006, 25:206-
      213.

[25] McCabe DE, Swain WF, Martinell BJ, Christou P. Stable transformation of soybean
      (Glycine max) by particle acceleration. Biotechnology 1988, 6:923-926.

[26] Saka H, Voqui Dinh TH, Cheng TY. Stimulation of multiple shoot formation on soy-
      bean stem nodes in culture. Plant Science Letters 1980, 19 193-201.

[27] Kim JH, Lamotte E Hack E. Plant regeneration in vitro from primary leaf nodes of
      soybean (Glycine max) seedling. Journal Plant Physiology 1990, 136:664-669.

[28] Wright MS, Williams MH, Pierson PE, Carnes MG. Initiation and propagation of Gly-
      cine max L. Merrill; Plants from tissue-cultured epicotyls. Plant Cell Tissue Organ
      Culture 1987, 8:83-90.

[29] Wright MS, Ward DV, Hinchee MA, Carnes MG, Kaufman RJ. Regeneration of soy-
      bean (Glycine max L. Merr.) from cultured primary leaf tissue. Plant Cell Reports
      1987, 6:83-89.

[30] Mante S, Scorza R, Cordts J. A simple, rapid protocol for adventitious shoot develop-
      ment from mature cotyledons of Glycine max cv.Bragg. In Vitro Cellular and Devel-
      opmental Biology Plant 1989, 25(4):385-388.

[31] Franklin G, Carpenter L, Davis E, Reddy CS, Al-Abed D, Alaiwi WA, Parani M,
      Smith B, Goldman SL, Sairam RV. Factors influencing regeneration of soybean from
      mature and immature cotyledons. Plant Growth Regulation 2004, 43:73-79.

[32] Yang N and Christou P. Cell type specific expression of a CaMV 35s-GUS gene in
      transgenic soybean plants. Developmental Genetics 1990, 11:289-293.

[33] Dan, Y, Reichert NA. Organogenic regeneration of soybean from hypocotyl explants.
      In Vitro Cellular and Developmental Biology Plant 1998, 34 14-21.

[34] Yoshida T. Adventitious shoot formation from hypocotyl sections of mature soybean
      seeds. Breeding Science 2002, 52:1-8.

[35] Liu HK, Yang C, Wie ZM. Efficient *Agrobacterium tumefaciens* mediated transforma-
      tion of soybeans using an embryonic Tip regeneration system. Planta 2004,
      219:1042-1049.

[36] Ma XH, Wu TL. Rapid and efficient regeneration in soybean [Glycine max (L.) Mer-
      rill] from whole cotyledonary node explants. Acta Physiologiae Plantarum 2008,
      30:209-216.

[37] Murashige T. and Skoog F. A revised medium for rapid growth and bioassays with tobacco tissue cultures. Physiologia Plantarum 1962, 15:473-479.

[38] Gamborg OL, Miller RA, Ojima K. Nutrient requirements of suspension cultures of soybean root cells. Experimental Cell Research 1968, 50:151-158.

[39] Joyner EY, Boykin LS, Lodhi MA. Callus Induction and Organogenesis in Soybean [Glycine max (L.) Merr.] cv. Pyramid from Mature Cotyledons and Embryos. The Open Plant Science Journal 2010, 4:18-21.

[40] Kim J, Hack E, LaMotte CE. Synergistic effects of proline and inorganic micronutrients and effects of individual micronutrients on soybean (Glycine max) shoot regeneration in vitro. Journal of Plant Physiology 1994, 144:726-734.

[41] Reichert NA, Young MM, Woods AL. Adventitious organogenic regeneration from soybean genotypes representing nine maturity groups. Plant Cell, Tissue and Organ Culture 2003, 75:273-277.

[42] Hu CY, Wang L. In planta soybean trasnformation technologies developed in China: procedure, confirmation and field performance. In Vitro Cellular and Developmental Biology Plant 1999, 35:417-420.

[43] Hong HP, Zhang H, Olhoft P, Hill S, Wiley H, Toren E, Hillebrand H, Jones T, Cheng M. Organogenic callus as the target for plant regeneration and transformation via Agrobacterium in soybean (Glycine max (L) Merr). In Vitro Cellular and Developmental Biology Plant, 2007, 43(6):558-568.

[44] Delzer BW, Somers DA, Orf JH. Agrobacterium tumefaciens susceptibility and plant regeneration of 10 soybean genotypes in maturity groups 00 to II. Crop Science 1990, 30:320-322.

[45] Chee PP, Fober KA, Slightom JL. Transformation of soybean (Glycine max) by infecting germinating seeds with Agrobacterium tumefaciens. Plant Physiology 1989, 91:1212-1218.

[46] Stewart CN Jr, Adang MJ, All JN, Barmy HR, Cardineau G, Tucker D, Parrot WA. Genetic transformation, recovery and characterization of fertile soybean [Glycine max (L.) Merr.] transgenic for a synthetic Bacillus thuringensis cryIAc gene. Plant Physiology 1995, 112:121-129.

[47] Aragao FJL, Sarokin L, Vianna GR, Rech, EL. Selection of transgenic meristematic cells utilizing a herbicidal molecule results in the recovery of fertile transgenic soybean [Glycine max (L) Merrill] plants at a high frequency. Theoretical and Applied Genetics 2000, 101 (1&2):1-6.

[48] Santarem ER Trick HN Essig JS Finer JJ. Sonication-assisted Agrobacterium-mediated transformation of soybean immature cotyledons: optimization of transient expression. Plant Cell Reports 1998, 17:752-759.

[49] Zhang Z, Guo Z, Shou H, Pegg SE, Clemente TE, Staswick PE, Wang K. Assessment of conditions affecting Agrobacterium-mediated soybean transformation and routine

recovery of transgenic soybean. In: A.D. Arencibia (Ed.), Plant Genetic Engineering: Towards the Third Millennium: Proceedings of the International Symposium on Plant Genetic Engineering–10 December 1999, 88-94. Havana, Cuba, Elsevier, Amsterdam / New York 2000.

[50] Meurer CA, Dinkins RD, Collins GB. Factors affecting soybean cotyledonary node transformation. Plant Cell Reports 1998, 18:180-186.

[51] Donaldson PA, Simmonds DH. Susceptibility of Agrobacterium tumefaciens and cotyledonary node transformation in short-season soybean. Plant Cell Reports 2000, 19:478-484.

[52] Maruyama EK, Ishi A, Migita S, Migita K. Screening of suitable sterilization of explants and proper media for tissue culture of eleven tree species of Peru-Amazon forest. Journal of Agricultural Science 1989, 33:252-261.

[53] Perl A, Lotan O, Abu-Abied M, Holland D. Establishment of an Agrobacterium-mediated transformation system for grape (Vitis vinifera L.): the role of antioxidants during grape-Agrobacterium interactions. Nature Biotechnology 1996, 14(5):624-628.

[54] Enriquez-Obregon GA, Prieto-Samsonov DL, de la Riva GA, Perez M, Selman-Housein G, Vasquez-Padron RI. Agrobacterium-mediated Japonica rice transformation: a procedure assisted by antinecrotic treatment. Plant Cell, Tissue and Organ Culture 1999, 59:159-168.

[55] Frame BR, Shou V Chikwamba RK, Zhang Z, Xiang C, Fonger TM, Pegg SEK, Li B, Nettleton DS, Pei D, Wang K. Agrobacterium tumefaciens-mediated transformation of maize embryos using standard binary vector system. Plant Physiology 2002, 129:13-22.

[56] Olhoft PM, Lin K, Galbraith J, Nielsen NC, Somers DA. The role of thiol compounds increasing Agrobacterium-mediated transformation of soybean cotyledonary-node cells. Plant Cell Reports 2001, 20:731-737.

[57] Olhoft PM, Flagel LE, Donovan CM, Somers DA. Efficient soybean transformation using hygromycin B selection in the cotyledonary-node method. Planta 2003, 216:723-735.

[58] Yun CS. High-efficiency Agrobacterium-mediated Transformation of Soybean. Acta Botanica Sinica 2004, 46 (5):610-617.

[59] Olhoft PM, Bernal LM, Grist LB, Hill DS, Mankin SL, Shen YW, Kalogerakis M, Wiley H, Toren E, Song HS, Hillebrand H, Jones T. A novel Agrobacterium rhizogenes-mediated transformation method of soybean [Glycine max (L.) Merrill] using primary-node explants from seedlings. In Vitro Cellular and Developmental Biology Plant 2007, 43:536-549.

[60] Clemente TE, LaVallee BJ, Howe AR, Conner Ward D, Rozman R, Hunter P, Broyles DL, Kasten D, Hinchee MA. Progeny analysis of glyphosate selected transgenic soy-

beans derived from Agrobacterium-mediated transformation. Crop Science 2000, 40:797-803.

[61] Zhang Z, Xing A, Staswick PE, Clemente TE. The use of glufosinate as a selective agent in Agrobacterium-mediated transformation of soybean. Plant Cell Tissue Organ Culture 1999, 56:37-46.

[62] Christou P, Swain WF, Yang NS, McCabe DE. Inheritance and expression of foreign genes in transgenic soybean plants. Proceedings of the National Academy of Sciences of the United States of America 1989, 86:7500-7504.

[63] Christou P, McCabe DE, Martinell BJ, Swain WF. Soybean genetic engineering-commercial products of transgenic plants. Trends in Biotechnology 1990, 8:145-15.

[64] Finer JJ, McMullen MD. Transformation of soybean via particle bombardment of embryogenic suspension culture tissue. In Vitro Cellular and Developmental Biology Plant 1991. 27:175-182.

[65] Donaldson PA, Anderson T, Lane BG, Davidson AL, Simmonds DH. Soybean plants expressing an active oligomeric oxalate oxidase from the wheat gf-2.8 (germin) gene are resistant to the oxalate-secreting pathogen Sclerotina sclerotiorum. Physiol. Mol. Plant Path 2001, 59:297-307.

[66] Piller KJ, Clemente TE, Jun SM, Petty CC, Sato S, Pascual DW, Bost KL. Expression and immunogenicity of an Escherichia coli K99 fimbriae subunit antigen in soybean. Planta 2005, 222:6-18.

[67] Xue RG, Zhang B, Xie HF. Overexpression of a NTR1 in transgenic soybean confers tolerance to water stress. Plant Cell Tissue Organ Culture 2007, 89:177-183.

[68] Chen DF, Zhang M, Wang YQ, Chen XW. Expression of γ-tocopherol methyltransferase gene from Brassica napus increased α-tocopherol content in soybean seed. Biologia Plantarum 2012, 56(1):131-134.

[69] Jiang N, Jeon EH, Pak JH, Ha TJ, Baek IY, Jung WS, Lee JH, Kim DH, Choi HK, Cui Z, Chung YS. Increase of isoflavones in soybean callus by Agrobacterium-mediated transformation. Plant Biotechnology Reporter 2010, 4 253-260.

[70] Yang S, Li G, Li M, Wang J. Transgenic soybean with low phytate content constructed by Agrobacterium transformation and pollen-tube pathway. Euphytica 2011, 177:375-382.

[71] Merkle SA, Parrott WA and Flinn BS. Morphogenic aspects of somatic embryogenesis. In: Thorpe TA (ed.): In vitro Embryogenesis in Plants. Kluwer Academic Publishers 1995, 155-203.

[72] McKersie BD and Brown DCW. Somatic embryogenesis and artificial seeds in Forage legumes. Seed Science Research 1996, 6:109-126.

[73] Reinert J. Morphogenese und ihre kontrolle an gewebekulturen aus carotten. Naturwissenchaften 1958, 45:344-345.

[74] Ammirato PV. The regulation of somatic embryo development in plant cell culture: suspension culture, techniques and hormone requirements. Nature Biotechnology 1983, 1:68-74.

[75] Mathews HR, Litz H, Wilde S, Merkle H and Wetzstein HY. Stable gene expression of β-glcuronidase and npt II genes in mango somatic embryos. In vitro Cellular and Developmental Biology Plant 1992, 28:172-178.

[76] Beversdorf WD, Bingham ET. Degrees of differentiation obtained in tissue cultures of Glycine species. Crop Science 1977, 17(2):307-311.

[77] Christianson ML, Warick DA, Carlson PS. A morphogenetically competent soybean suspension culture. Science 1983, 222:632-634.

[78] Von AS, Sabala I, Bozhkov P, Dyachok J, Filonova L. Developmental pathways of somatic embryogenesis. Plant Cell Tissue Organ Culture 2002, 69:233-249

[79] Finer JJ. Apical proliferation of embryogenic tissue of soybean (Glycine max (L.) Merrill). Plant Cell Reports 1988, 7(4):238-241.

[80] Hoffmann N, Nelson RL, Korban SS. Influence of media components and pH on somatic embryo induction in three genotypes of soybean. Plant Cell, Tissue and Organ Culture 2004, 77(2):157-163.

[81] Santarem ER, Pelissier B, Finer JJ. Effect of explant orientation, pH, solidifying agent and wounding on initiation of soybean somatic embryos. In Vitro Cellular and Developmental Biology-Plant 1997, 33 (1):13-19.

[82] Finer JJ, Nagasawa A. Development of an embryogenic suspension culture of soybean (Glycine max Merrill). Plant Cell, Tissue and Organ Culture 1988, 15:125-136.

[83] Liu W, Moore PJ, Collins GB. Somatic embryogenesis in soybean via somatic embryo cycling. In Vitro Cellular and Developmental Biology Plant 1992, 28:153-160.

[84] Lippmann B, Lippmann G. Induction of somatic embryos in cotyledonary tissue of soybean, Glycine max L. Merr. Plant Cell Reports 1984, 185(3):215-218.

[85] Lazzeri PA, Hildebrand DF, Collins GB. Soybean somatic embryogenesis: effects of hormones and culture manipulations. Plant Cell, Tissue and Organ Culture 1987, 10:197-208.

[86] Lazzeri PA, Hildebrand DF, Sunega J, Williams EG, Collins GB. Soybean somatic embryogenesis: interactions between sucrose and auxin. Plant Cell Reports 1988, 7:517-520.

[87] Parrott WA, Dryden G, Vogt S, Hildebrand DF, Collins GB, Williams EG. Optimization of somatic embryogenesis and embryo germination in soybean. In Vitro Cellular and Development Biology Plant 1988, 24:817-820.

[88] Moon H, Hildebrand DF. Effects of proliferation, maturation, and desiccation methods on conversion of soybean somatic embryos. In Vitro Cellular and Developmental Biology Plant 2003, 39:623-628

[89] Samoylov VM, Tucker DM, Parrott WA. Soybean [Glycine max (L.) Merrill] embryogenic cultures: the role of sucrose and total nitrogen content on proliferation. In Vitro Cellular and Developmental Biology Plant 1998, 34:8-13.

[90] Walker DR, Parrott WA. Effect of polyethylene glycol and sugar alcohols on soybean somatic embryo germination and conversion. Plant Cell, Tissue and Organ Culture 2001, 64(1):55-62.

[91] Korbes AP, Droste A. Carbon sources and polyethylene glycol on soybean somatic embryo conversion. Pesquisa Agropecuária Brasileira 2005, 40(3):211-216.

[92] Santos KGB, Mundstock E, Bodanese Zanettini MH. Genotype-specific normalization of soybean somatic embryogenesis through the use of an ethylene inhibitor. Plant Cell Reports 1997, 16 (12):859-864.

[93] Jang GW, Park RD, Kim KS. Plant regeneration from embryogenic suspension cultures of soybean (Glycine max L. Merrill). Journal of Plant Biotechnology 2001, 3:101-106.

[94] Buchheim JA, Colburn SM, Ranch JP. Maturation of soybean somatic embryos and the transition to plantlet grown. Plant Physiology 1989, 89:768-77.

[95] Durham RE and Parrott WA. Repetitive somatic embryogenesis from peanut cultures in liquid medium. Plant Cell Reports 1992, 11:122-125.

[96] Bailey MA, Boerma HR, Parrott WA. Genotype effects on proliferative embryogenesis and plant regeneration of soybean. In vitro Cellular and Developmental Biology Plant 1993, 29:102-108.

[97] Simmonds DH, Donaldson PA. Genotype screening for proliferative embryogenesis and biolistic transformation of short-season soybean genotypes. Plant Cell Reports 2000, 19:485-490.

[98] Yang C, Zhao T, Yu D, Gai J. Somatic embryogenesis and plant regeneration in Chinese soybean (Glycine max (L.) Merr.) Impacts of mannitol, abscisic acid, and explants age. In Vitro Cellular and Developmental Biology-Plant 2009, 45 (2):180-181.

[99] Bonacin GA, DiMauro AO, Oliveira RC, Perecin D. Induction of somatic embryogenesis in soybean: physicochemical factors influencing the development of somatic embryos. Genetics and Molecular Biology 2000, 23(4):865-868.

[100] Droste A, Pasquali G, Bodanese-Zanettini, MH. Transgenic fertile plants of soybean [Glycine max (L) Merrill] obtained from bombarded embryogenic tissue. Euphytica 2002, 127(3):367-376.

[101] Bailey MA, Boerma HR, Parrott WA. Genotype-specific optimization of plant regeneration from somatic embryos of soybean. Plant Science 1993, 93:117-120.

[102] Hiraga S, Minakawa H, Takahashi K, Takahashi R, Hajika M, Harada K, Ohtsubo N. Evaluation of somatic embryogenesis from immature cotyledons of Japanese soybean cultivars. Plant Biotechnology 2007, 24 (4):435-440.

[103] Meurer CA, Dinkins RD, Redmond CT, Mcallister KP, Tucker DT, Walker DR, Parrott WA, Trick HN Essig JS, Frantz HM, Finer JJ, Collins GB. Embryogenic response of multiple soybean [Glycine max (L.) Merr.] cultivars across three locations. In Vitro Cellular and Developmental Biology Plant 2001, 37(67):62-67.

[104] Kita Y, Nishizawa K, Takahashi M, Kitayama M, Ishimoto M. Genetic improvement of the somatic embryogenesis and regeneration in soybean and transformation of the improved breeding lines. Plant Cell Reports 2007, 26(4):439-447.

[105] Parrott WA, Hoffman LM, Hildebrand DF, Williams EG, Collins GB. Recovery of primary transformants of soybean. Plant Cell Reports 1989, 7:615-61.

[106] Hood EE, Gelvin SB, Melchers LS and Hoekema A. New Agrobacterium helper strains for gene transfer to plants. Transgenic Research 1993, 2:208-218.

[107] Torisky RS, Kovacs L, Avdiushko S, Newman JD, Hunt AG and Collins GB. Development of a binary vector system for plant transformation based on the supervirulent Agrobacterium tumefaciens strain Chry5. Plant Cell Reports 1997, 17:102-108.

[108] Trick HN, Finer JJ. Sonication-assisted Agrobacterium-mediated transformation of soybean (Glycine max) embryogenic suspension culture tissue. Plant Cell Reports 1998, 17:482-488.

[109] Yan B, Reddy MSS, Collins GB, Dinkins RD. *Agrobacterium tumefaciens* mediated transformation of soybean [Glycine max (L) Merrill] using immature zygotic cotyledon explants. Plant Cell Reports 2000, 19(11):1090-1097.

[110] Ko TS, Korban SS. Enhancing the frequency of somatic embryogenesis following Agrobacterium-mediated transformation of immature cotyledons of soybean [Glycine max (L.) Merrill.]. In Vitro Cellular & Developmental BiologyPlant 2004, 40:552-558.

[111] Ko TS, Lee S, Farrand SK, Korban SS. A partially disarmed vir helper plasmid, pKYRT1, in conjunction with 2,4-dichlorophenoxyacetic acid promotes emergence of regenerable transgenic somatic embryos from immature cotyledons of soybean. Planta 2004, 218(4):536-541.

[112] Wiebke Strohm B, Droste A, Pasquali G, Osorio MB, Bucker Neto L, Passaglia LMP, Bencke M, Homrich MS, Margis Pinheiro M, Bodanese Zanettini MH. Transgenic fertile soybean plants derived from somatic embryos transformed via the combined DNA-free particle bombardment and Agrobacterium system. Euphytica 2011, 177(3): 343-354.

[113] Hazel CB, Klein TM, Anis M, Wilde HD, Parrott WA. Growth characteristics and transformability of soybean embryogenic cultures. Plant Cell Reports 1998, 17:765-772.

[114] Maughan PJ, Philip R, Cho MJ, Widholm JM and Vodkin LO. Biolistic transformation, expression, and inheritance of bovine β-casein in soybean (Glycine max). In vitro Cellular and Developmental Biology Plant 1999, 35:334-349.

[115] Hadi MZ, McMullen MD, Finer JJ. Transformation of 12 different plasmids into soybean via particle bombardment. Plant Cell Reports 1996, 15:500-505.

[116] Santarem ER, Finer JJ. Transformation of soybean [Glycine max (L.) Merrill] using proliferative embryogenic tissue maintained on semisolid medium. In Vitro Cellular and Developmental Biology Plant 1999, 35 451-455.

[117] Dufourmantel N, Pelissier B, Garcon F, Peltier G, Jean-Marc F, Tissot G. Generation of fertile transplastomic soybean. Plant Molecular Biology 2004, 55 479-489.

[118] Rao SS, Hildebrand D. Changes in Oil Content of Transgenic Soybeans Expressing the Yeast SLC1 Gene. Lipids 2009, 44:945-951.

[119] Chen R, Matsui K, Ogawa M, Oe M, Ochiai M, Kawashima H, Sakuradani E, Shimizu S, Ishimoto M, Hayashi M, Murooka Y, Tanaka Y. Expression of Delta 6, Delta 5 desaturase and GLELO elongase genes from Mortierella alpina for production of arachidonic acid in soybean [Glycine max (L.) Merrill] seeds. Plant Science 2006, 170:399-406.

[120] Tougou M, Furutani N, Yamagishi N, Shizukawa Y, Takahata Y, Hidaka S. Development of resistant transgenic soybeans with inverted repeat-coat protein genes of soybean dwarf virus. Plant Cell Reports 2006, 25:1213-1218.

[121] Furutani N, Yamagishi N, Hidaka S, Shizukawa Y, Kanematsu S, Kosaka Y. Soybean mosaic virus resistance in transgenic soybean caused by post-transcriptional gene silencing. Breed Science 2007, 57:123-128.

[122] Ishimoto M, Rahman SM, Hanafy MS, Khalafalla MM, El-Shemy HA, Nakamoto Y, Kita Y, Takanashi K, Matsuda F. Murano Y, Funabashi T, Miyagawa H, Wakasa K. Evaluation of amino acid content and nutritional quality of transgenic soybean seeds with high-level tryptophan accumulation. Molecular Breeding 2010, 25:313-326.

[123] Tavva VS, Kim YH, Kagan IA, Dinkins RD, Kim KH and Collins GB. Increased α-tocopherol content in soybean seed over- expressing the Perilla frutescens γ-tocopherol methyltransferase gene. Plant Cell Reports 2007, 26:61-70.

[124] Li Z, Meyer S, Essig JS, Liu Y, Schapaugh MA, Muthukrishnan S, Hainline BE,. Trick HN. High-level expression of maize γ-zein protein in transgenic soybean (Glycine max) Molecular Breeding 2005, 16:11-20.

[125] Dinkins RD, Srinivasa Reddy MS, Meurer CA, Yan B, Trick HN, Finer JJ, Thibaud-Nissen F, Parrott WA and Collins GB. Increased sulfur amino acids in soybean plants overexpressing the maize 15 kDa zein protein. In vitro Cellular and Developmental Biology Plant 2001, 37:742-747.

[126] Schnell JA, Treyvaud-Amiguet V, Arnason JT, Johnson DA. Expression of polyhy-
       droxybutyric acid as a model for metabolic engineering of soybean seed coats. Trans-
       genic Research 2012, 21(4):895-899.

[127] Luo XM, Zhao GL, Jian YY. Plant regeneration from protoplasts of soybean (Glycine
       max L.). Acta Botanica Sinica 1990, 32:616-621.

[128] Yeh MS. In vitro culture of immature soybean embryos II. The abilities of organogen-
       esis and plantlet regeneration from different aged immature embryo in Glycine spe-
       cies. Journal of the Agricultural Association of China 1990, 39:73-87.

[129] Dhir SK, Dhir S, Widholm JM. Plantlet regeneration from immature cotyledon proto-
       plasts of soybean (Glycine max L.). Plant Cell Reports 1991, 10:39-43.

[130] Sato S, Newell C, Kolacz K, Tredo L, Finer JJ, Hinchee M. Stable transformation via
       particle bombardment in two different soybean regeneration systems. Plant Cell Re-
       ports 1993, 12(7-8):408- 413.

[131] Di R, Purcell V, Collins GB, Ghabrial SA. Production of transgenic soybean lines ex-
       pressing the bean pod mottle virus coat protein precursor gene. Plant Cell Reports
       1996, 15:746-750.

[132] Xing A, Zhang Z, Sato S, Staswick PE, Clemente TE. The use of the two T-DNA bina-
       ry system to derive marker-free transgenic soybeans. In Vitro Cellular and Develop-
       mental Biology Plant 2000, 36:456-463.

[133] Wang X, Eggenberger AL, Nutter Jr FW, Hill JH. Pathogen-derived transgenic resist-
       ance to soybean mosaic virus in soybean. Molecular Breeding 2001, 8:119-127.

[134] Zeng P, Vadnais DA, Zhang Z, Polacco JC. Refined glufosinate selection in Agrobac-
       terium-mediated transformation of soybean [Glycine max (L.) Merrill]. Plant Cell Re-
       ports 2004. 22:478-482.

[135] Miklos JA, Alibhai MF, Bledig SA, Connor Ward DC, Gao AG, Holmes BA, Kolacz
       KH, Kabuye VT, MacRae TC, Paradise MS, Toedebusch AS, Harrison LA. Characteri-
       zation of soybean exhibiting high expression of a synthetic Bacillus thuringiensis
       cry1A transgene that confers a high degree of resistance to Lepidopteran pests. Crop
       Science 2007, 47:148-157.

[136] Liu SJ Wei ZM, Huang JQ. The effect of co-cultivation and selection parameters on
       Agrobacterium-mediated transformation of Chinese soybean varieties. Plant Cell Re-
       ports 2008, 27(3):489-498.

[137] Lee K, Yi BY, Kim KH, Kim JB, Suh SC, Woo HJ, Shin KS, Kweon SJ. Development of
       Efficient Transformation Protocol for Soybean (Glycine max L.) and Characterization
       of Transgene Expression after Agrobacterium-mediated Gene Transfer. Journal of the
       Korean Society for Applied Biological Chemistry 2011, 54:37-45.

[138] Lazzeri PA, Hilderbrand DF, Collins GB. A procedure for plant regeneration from immature cotyledon tissue of soybean. Plant Molecular Biology Reporter 1985, 3(4): 160-167.

[139] Ranch JP, Ogelsby L, Zielinski AC. Plant regeneration from embryo-derived tissue cultures of soybean. In Vitro Cellular and Developmental Biology Plant 1985, 21:653-658.

[140] Ghazi TD, Cheema HV, Nabors MW. Somatic embryogenesis and plant regeneration from embryonic callus of soybean [Glycine max (L.) Merr.]. Plant Cell Reports 1986, 5(6):452-456.

[141] Kerns HR, Barwale VB, Meyer MM. Correlation of cotyledonary node shoot proliferation and somatic embryoid development in suspension cultures of soybean [Glycine max (L.) Merr.]. Plant Cell Reports 1986, 5(2), 140-143.

[142] Hartweck LM, Lazzeri PA, Cui D, Collins GB, Williams EG. Auxin orientation effects on somatic embryogenesis from immature soybean cotyledons. In Vitro Cellular and Developmental Biology Plant 1988, 24(8):821-828.

[143] Komatsuda T, Ohyama K. Genotype of high competence for somatic embryogenesis and plant regeneration in soybean Glycine max. Theoretical and Applied Genetics 1988, 75(5):695-700.

[144] Parrott WA, Williams EG, Hildebrand DF, Collins GB. Effect of genotype on somatic embryogenesis from immature cotyledons of soybean. Plant Cell, Tissue and Organ Culture 1989, 16(1):15-21.

[145] Christou, P, Yang NS. Developmental aspects of soybean (Glycine max) somatic embryogenesis. Annals of Botany 1989, 64(2):225-234.

[146] Komatsuda T, Ko SW. Screening of soybean (Glycine max (L.) Merrill) genotypes for somatic embryo production from immature embryo. Japanese Journal of Breeding 1990, 40:249-251.

[147] Komatsuda T, Kanebo K, Oka S. Genotype × sucrose interactions for somatic embryogenesis in soybean. Crop Science 1991, 31(2):333-337.

[148] Wright MS, Launis KL, Novitzky R, Duesiing JH, Harms CT. A simple method for the recovery of multiple fertile plants from individual somatic embryos of soybean [Glycine max (L.) Merrill]. In Vitro Cellular and Developmental Biology Plant 1991, 27:153-157.

[149] Lippmann B, Lippmann G. Soybean embryo culture: factors influencing plant recovery from isolated embryos. Plant Cell, Tissue and Organ Culture 1993, 32(1):83-90.

[150] Parrott WA, All JN, Adang MJ, Bailey MA, Boerma HR, Stewart CN Jr. Recovery and evaluation of soybean plants transgenic for a Bacillus thuringiensis var.kurstaki insecticidal gene. In Vitro Cellular and Development Biology Plant 1994, 30:144-149.

[151] Saravitz CH, Raper CDJr. Responses to sucrose and glutamine by soybean embryos grown in vitro. Physiologia Plantarum 1995, 93:799-805.

[152] Liu W, Torisky RS, McAllister KP, Avdiushko S, Hildebrand D, Collins GB. Somatic embryo cycling: evaluation of a novel transformation and assay system for seed-specific gene expression in soybean. Plant Cell, Tissue and Organ Culture 1996, 47:33-42.

[153] Rajasekaran K, Pellow JW. Somatic embryogenesis from cultured epicotyls and primary leaves of soybean [Glycine max (L.) Merrill]. In Vitro Cellular and Developmental Biology Plant 1997, 33:88-91.

[154] Samoylov VM, Tucker DM, Parrott WA. A liquid medium-based rapid regeneration from embryogenic soybean cultures. Plant Cell Reports 1998, 18 49-54.

[155] Chanprame S, Kuo TM, Widholm AM. Soluble carbohydrate content of soybean [Glycine max (L.) Merr.] somatic and zygotic embryos during development. In Vitro Cellular and Developmental Biology Plant 1998, 34:64-68.

[156] Ponappa T, Brzozowski AE, Finer JJ. Transient expression and stable transformation of soybean using jellyfish green fluorescent protein. Plant Cell Reports 1999, 19:6-12.

[157] Tian L, Brown DCW. Improvement of soybean somatic embryo development and maturation by abscisic acid treatment. Canadian Journal of Plant Science 2000. 80:721-276.

[158] Walker DR, All JN, McPherson RM, Boerma HR, Parrott WA. Field Evaluation of Soybean Engineered with a Synthetic cry1Ac Transgene for Resistance to Corn Earworm, Soybean Looper, Velvetbean Caterpillar (Lepidoptera: Noctuidae), and Lesser Cornstalk Borer (Lepidoptera: Pyralidae). Journal of Economic Entomology, 2000, 93(3) 613-622.

[159] Droste A, Pasquali G, Bodanese-Zanettini MH. Integrated bombardment and Agrobacterium transformation system: an alternative method for soybean transformation. *Plant Molecular Biology Reporter* 2000, 18:51-59.

[160] Droste A, Leite PCP, Pasquali G, Mundstock EC, Bodanese-Zanettini MH. Regeneration of soybean via embryogenic suspension culture. Scientia Agricola 2001, 58(4): 753-758.

[161] Fernando JA, Vieira MLC, Geraldi IO, Appezzato-da-Gloria B. Anatomical study of somatic embryogenesis in Glycine max (L.) Merrill. Brazilian Archives of Biology and Technology 2002, 45 (3):277-286.

[162] Tomlin ES, Branch SR, Chamberlain D, Gabe H, Wright MS, Stewart CN Jr. Screening of soybean, Glycine max (L.) Merrill, lines for somatic embryo induction and maturation capability from immature cotyledons. In Vitro Cellular and Developmental Biology Plant 2002, 38:543-548.

[163] Chiera JM, Finer JJ, Grabau EA. Ectopic expression of a soybean phytase in developing seeds of Glycine max to improve phosphorus availability. Plant Molecular Biology 2004, 56:895-904.

[164] Schmidt MA, Tucker DM, Cahoon EB, Parrott WA. Towards normalization of soybean somatic embryo maturation. Plant Cell Reports 2005, 24:383- 391.

[165] Santos KGB, Mariath JEA, Moco MCC, Bodanese Zanettini MH. Somatic Embryogenesis from Immature Cotyledons of Soybean (Glycine max (L.) Merr.): Ontogeny of Somatic Embryos. Brazilian Archives of Biology and Technology 2006, 49 (1):49-55.

[166] Wiebke B, Ferreira F, Pasquali G, Bodanese Zanettini MH, Droste A. Influence of antibiotics on embryogenic tissue and Agrobacterium tumefaciens suppression in soybean genetic transformation. Bragantia 2006, 65(4):543-551.

[167] Chiera JM, Streeter JG, Finer JJ. Ononitol and pinitol production in transgenic soybean containing the inositol methyl transferase gene from Mesembryanthemum crystallinum. Plant Science 2006. 171:647-654.

[168] Tougou M, Yamagishi N, Furutani N, Shizukawa Y, Takahata Y, Hidaka S. Soybean dwarf virus-resistant transgenic soybeans with the sense coat protein gene. Plant Cell Reports 2007, 26:1967-1975.

[169] El-Shemy H A, Khalafalla MM, Fujita K, Ishimoto M. Improvement of protein quality in transgenic soybean plants. Biologia Plantarum 2007, 51:277-284.

[170] Weber RLM, Korber AP, Baldasso DA, Callegari Jacques SM, Bodanese Zanettini MH, Droste A. Beneficial effect of abscisic acid on soybean somatic embryo maturation and conversion into plants. Plant Cell Culture and Micropropagation 2007, 3(1): 1-9.

[171] Yemets AI, Radchuk VV, Pakhomov AV, Blume Ya B. Biolistic Transformation of Soybean Using a New Selectable Marker Gene Conferring Resistance to Dinitroanilines. Cytology and Genetics 2008, 42(6):413-419.

[172] Schmidt MA, LaFayette PR, Artelt BA, Parrott WA. A comparison of strategies for transformation with multiple genes via microprojectile-mediated bombardment. In Vitro Cellular and Developmental Biology Plant 2008, 44(3):162-168.

[173] Loganathan M Maruthasalam S, Shiu LY, Lien WC, Hsu WH, Lee PF, Yu CW, Lin CH. Regeneration of soybean (Glycine max L. Merrill) through direct somatic embryogenesis from the immature embryonic shoot tip. In Vitro Cellular and Developmental Biology Plant 2010, 46:265-273.

[174] Droste A, Silva da AM, Souza de IF, Wiebke-Strohm, Bucker Neto L, Bencke M, Sauner MV, Bodanese-Zanettin MH. Screening of Brazilian soybean genotypes with high potential for somatic embryogenesis and plant regeneration. Pesquisa Agropecuaria Brasileira 2010, 45(7):715-720.

[175] He Y, Young TE, Clark KR, Kleppinger-Sparace K F, Bridges WC, Sparace SA. Developmental profile of storage reserve accumulation in soybean somatic embryos. In Vitro Cellular and Developmental Biology Plant 2011, 47:725-733.

[176] Mariashibu TS, Subramanyam K, Arun M, Mayavan S, Rajesh M, Theboral J, Manickavasagam M, Ganapathi A. Vacuum infiltration enhances the Agrobacterium-mediated genetic transformation in Indian soybean cultivars. Acta Physiologiae Plantarum 2012, (published online) DOI 10.1007/s11738-012-1046-3.

Chapter 2

Soybean Proteomics: Applications and Challenges

Alka Dwevedi and Arvind M Kayastha

Additional information is available at the end of the chapter

1. Introduction

Proteomics is one of the most explored areas of research based on global-scale analysis of proteins. It leads to direct understanding of function and regulation of genes. Significant advances in the comprehensive profiling, functional analysis, and regulation of plant proteins have not advanced much as compared to model organisms such as yeast, humans etc. The application of proteomic approaches to plants implicates; comprehensive identification of proteins, their isoforms, as well as their prevalence in each tissue, characterizing the biochemical and cellular functions of each protein and the analysis of protein regulation and its relation to other regulatory networks [1]. Genes of higher eukaryotes (including plants) contain introns which are large and numerous. Therefore, combinational exon usage originating from complex gene structures results in a multitude of splice variants leading to generation of different protein products from a given gene. Thus, the determination of the comprehensive pattern of expression of each protein isoform is a challenging task, most importantly for poorly expressed proteins [2].

The two-dimensional gel electrophoresis (2-DE) is used for profiling protein expression involving separation of complex protein mixtures by molecular charge in the first dimension and by mass in the second dimension. Recent advancement in 2-DE has improved resolution and reproducibility but still automation in high-throughput setting is lagging. The alternative approaches like multi-dimensional protein identification technology involving large-scale proteomics are able to generate a large catalog of proteins present in complex cell extracts. Further, detection of low abundance proteins using sub-cellular fractionation reduces the complexity of protein extracts. These efforts have successfully characterized nuclear, chloroplast, amyloplast, plasma membrane, peroxisome, endoplasmic reticulum, cell wall, and mitochondrial proteomes of a model plant, *Arabidopsis*. Although, high-throughput technologies have helped in characterization of *Arabidopsis* and other organisms' proteomes, characterization of various protein classes including membrane and hydrophobic proteins which are recalcitrant to isolation and analysis is still inaccessible [3].

Food allergy can be a serious nutritional problem in children and adults. Any protein-containing food has the potential to elicit an allergic reaction in the human population. Antibody IgE-mediated reactions are the most prevalent allergic reactions to food. These responses occur after the release of chemical mediators from mast cells and basophils as a result of interactions between food proteins and specific IgE molecules on the surface of these receptor cells. Eight foods or food groups have been identified as the most fre-quent sources of human food allergens and account for over 90% of the documented food allergies worldwide. These foods are milk, eggs, fish, crustaceans, wheat, peanuts, tree nuts and soy [4]. Despite their well-documented allergenicity, soy derivatives continue to be increasingly used in a variety of food products due to their well-documented health benefits. Soybean has also been one of the selected target crops for genetic modification (GM). For example, the artificial introduction of 5-enolpyruvylshikimate-3-phosphate syn-thase in soybean crop creates an alternative pathway which is insensitive to glyphosate (most potent herbicide), thus increasing overall crop yield. One of the major concerns regarding the safety of GM foods is the potential allergenicity of the resulting products, namely the possible occurrence of either altered or *de novo* expressed of endogenous aller-gens after genetic manipulation. This concern justifies careful plant characterization [5]. Proteomics is one of the powerful approaches allowing rapid and reliable protein identi-fication. It can provide information about their post-translational modifications, sub-cellu-lar localization, level of protein expression and protein-protein interactions. Despite the importance of soybean and the availability of powerful tools for the analysis of proteins from sub-cellular organelles, and specifically for the identification of allergens, only a lim-ited number of reports have been published to date.

Soybean is an important source of protein for human and animal nutrition, as well as a ma-jor source of vegetable oil. Although soybean is adapted to grow in a range of climatic con-ditions including adverse environmental and biological factors, still it has been affected with respect to growth, development, and global production For instance, drought reduces the yield of soybean by about 40%, affecting all stages of plant development from germination to flowering thus reducing the quality of the seeds. [6]. Several other abiotic stresses, such as flooding, high temperature, irradiation, or the presence of pollutants in the air and soil have detrimental effects on the growth and productivity of soybean. Along with morphological and physiological studies on the responses of plants to stress conditions, several molecular mechanisms from gene transcription to translation as well as metabolites were investigated. Recent advances in the field of proteomics have created an opportunity for dissecting quan-titative traits in a more meaningful way. Proteomics can investigate the molecular mecha-nisms of plants' responses to stresses and provides a path toward increasing the efficiency of indirect selection for inherited traits. In soybean a comprehensive functional genomics is yet to be performed; therefore, proteomics approaches form a powerful tool for analyzing the functions of complete set of proteins including those involved in stress protection.

2. Proteomics: isolation, identification and classification

In plant proteomics, the type of the plant species, tissues, organs, cell organelles, and the nature of desired proteins affect the techniques that can be used for protein extraction. Furthermore, the extraction process becomes more tedious when the protein is present inside vacuoles, rigid cell walls, or membrane plastids. A perfect protein extraction method involves complete solubilization of total proteins from a given sample and minimizing post-extraction artifact formation, proteolytic degradation as well as removal of non-proteinaceous contaminants. To date, only the proteome of *Arabidopsis* and rice have been studied while less attention has been paid to other plants including soybean. Soybean has high levels of phenolic compounds, proteolytic and oxidative enzymes, terpenes, organic acids, and carbohydrates due to which protein extraction is very tedious. Further it contains contains large quantities of secondary metabolites, *viz.* flavone glycosides (kaempferol and quercetin glycosides), phenolic compounds, lipids and carbohydrates. Thus impedes high-quality protein extraction in turn high-resolution protein separation in 2-DE.

In classical proteome analyses, proteins are initially separated by a 2-DE technique with iso-electric focusing (IEF) as the first dimension and sodium dodecyl sulfate polyacrylamide gel electrophoresis (SDS-PAGE) as the second dimension. A greater resolution in protein separation has been achieved by introducing immobilized pH gradients (IPGs) for the first dimension. Methodological advances in 2-DE have led to the introduction of two-dimensional fluorescence difference gel electrophoresis (2D-DIGE), which has been used for the comparative analysis of the proteome of soybean subjected to abiotic and biotic stresses [7]. The separated proteins can be subsequently identified by sequencing or by mass spectrometry. By introduction of mass spectrometry into protein chemistry, matrix-assisted laser desorption/ionization time-of-flight mass spectrometry (MALDI-TOF MS) and liquid chromatography/tandem mass spectrometry (LC-MS/MS) have become the methods of choice for high-throughput identification of proteins. An alternative technique known variously as 'gel-free proteomics', 'shotgun proteomics', or 'LC-MS/MS-based proteomics' can also be used in high-throughput protein analysis. This approach is based on LC separation of complex peptide mixtures coupled with tandem mass spectrometric analysis. A multidimensional protein identification technology (MudPIT) that usually incorporates separation on a strong cation exchange, reverse-phase column and MS/MS analysis helps the efficient separation of complex peptide mixtures. The gel-free technique have the advantage of being capable of identifying low-abundance proteins, proteins with extreme molecular weights or p*I* values, and hydrophobic proteins that cannot be identified by using gel-based technique. A combination of gel-based and gel-free proteomics has been used for identification of soybean plasma membrane proteins under abiotic stress, *viz.* flooding, osmotic, salinity stress. Methods for protein identification are not usually organism specific, and they can be applied to a wide range of living organisms in addition to soybean. Identification of proteins is normally performed by using a database search engine such as MASCOT or SEQUEST.

Soybean has an estimated genome size of 1115 Mbp, which is significantly larger than those of other crops, such as rice (490 Mbp) or sorghum (818 Mbp). Sequencing of the 1100 Mbp of total soybean genome predicts the presence of 46,430 protein-encoding genes, 70% more than in *Arabidopsis* [8]. The soybean genome database contains 75,778 sequences and 25,431,846 residues have been constructed on the basis of the Soybean Genome Project, DOE Joint Genome Institute; this database is available at http://www.phytozome.net. Although the genome sequence information is almost completed, no high-quality genome assembly is available because the results from the computational gene-modeling algorithm are imperfect. In addition, duplications in the genome of soybean result in nearly 75% of the genes being present as multiple copies, which further complicate the analysis. The soybean proteome database (http://proteome.dc.affrc.go.jp/Soybean) provides valuable information including 2-DE maps and functional analysis of soybean proteins. However, the presence of a considerable number of proteins with unknown functions highlights the limitations of bioinformatics prediction tools and the need for further functional analyses. The cellular proteomics helps in identification of changes in protein expression under different growing condition and treatments. The analytical methodology for the separation and identification of a large numbers of proteins should be authentic and confirmable. The proteome map of mature dry soybean seeds has been prepared by employing robotic automation at subsequent steps of 2-DE. Further, UniGene database was implemented for proteins identifications. Total protein from mature dry soybean (*Glycine max* cv. Jefferson) seed was isolated and 2D-PAGE performed using 13 cm IPG strips and subsequently doing SDS-PAGE. Protein spots were analyzed using Phoretix 2D-Advanced software. Excised protein spots were arrayed into 96-well plates and transferred to a Multiprobe II EX liquid handling station for subsequent destaining, tryptic digestion and peptide extraction. MALDI-TOF MS was operated in the positive ion delayed extraction reflector mode. Peptide spectra were submitted to a MS Fit program of Protein Prospector. Assignments from UniGene contigs were subsequently searched against the NCBI non-redundant database using the BLASTP search algorithm to determine similarity matches [9].

Trichloroacetic acid (TCA)/acetone-based and phenol-based buffers are most frequently used in protein extraction from plants. A comprehensive proteomic study was performed on nine organs from soybean plants in various developmental stages by using three different methods for protein extraction and solubilization. The results showed that the use of an alkaline phosphatase buffer followed by TCA/acetone precipitation caused horizontal streaking in 2-DE while use of a Mg/NP-40 buffer followed by extraction with alkaline phenol and methanol/ammonium acetate produced high-quality proteome maps with well-separated spots, high spot intensities, and high numbers of separate protein spots in 2-DE gels [10, 11]. In the case of organelle proteomics particularly that of membrane proteomics, a different extraction procedure is required that involves modifications to dissolve hydrophobic proteins and additional purification steps. Furthermore, when studying protein–protein interactions, it is necessary to extract protein complexes by using buffers with less or no detergent to get the proteins in their native states. Despite the importance of seed filling in the synthesis of storage reserves for germination, systematic proteomic analysis of this phase in legumes is yet to be carried out.

Total seed proteins of soybean (cv. Maverick) at different stages of flowering (14, 21, 28, 35 and 42 days) were isolated and subsequently 2D-PAGE was done. Initially IPG strips of pH 3 to 10 were taken then narrowed down to pH range to 4 to 7 for high-resolution proteome maps. A total of 488 and 679 proteins were identified from 2D-PAGE gels of pH range 4 to 7 and 3 to 10 gels, respectively. Each of the 679 proteins was excised from reference gels for identification by MALDI-TOF MS and a total of 422 proteins (62%) were identified. One unique protein was often represented by more than one spot on the 2D-PAGE gel, most likely due to post-translational modifications or genetic isoforms. Taking into account this redundancy, 216 unique proteins out of 422 were identified. A total of 82 proteins were associated with metabolism (the largest functional class) and the second largest functional class were comprised of 52 spots assigned to the seed storage proteins β-conglycinin and glycinin. An overall down- and up-regulation was observed for metabolism and storage related proteins, respectively, during seed filling, suggesting metabolic activity curtails as seeds approach maturity. Abundance of proteins related to metabolite transporter, disease and defense, energy production, cell growth and division, signal transduction, protein synthesis and secondary metabolism did not vary significantly. Furthermore, 13 sucrose-binding proteins have been mapped to the same UniGene accession number, suggesting the importance of sucrose as a signaling molecule in seed and embryo development. There were a total of 92 unknown proteins which could not be classified, therefore grouped into five expression profiles [12].

3. Implication of proteomics in understanding soybean stress

Soybean is grown worldwide with an average protein content of 40% (highest protein content with respect to other food crops) and oil content of 20% (which is second only to that of groundnut among the leguminous foods). Furthermore, soybean improves soil fertility by fixing nitrogen from the atmosphere in symbiosis with nitrogen fixing bacteria. It is, however, susceptible to various types of stresses (abiotic and biotic). Tolerance and susceptibility to stresses are complex phenomena because they are quantitatively inherited and can occur during different stages of plant growth and development. Extrinsic stress is regarded as the most important stress agent, which results from changes in abiotic factors such as temperature, climatic factors and chemical components, either naturally occurring or manmade. Further, biotic stresses (occurs as a result of damage done to plants by other living organisms, such as bacteria, viruses, fungi, parasites, beneficial and harmful insects, weeds, bacterial, fungal, algal and viral diseases) can also cause huge deterioration in plant growth and yield. Plants have developed adaptive features against these stresses. The genome remains unchanged to a large extent in any particular cell while proteins change dramatically as genes are turned on or off in response to stress. The proteome determines the cellular phenotype and its plasticity in response to external signals. It is proteins that are directly involved in both normal and stress-associated biochemical processes. Therefore, a more complete understanding of stress in soybean may be gained by looking directly into the proteins within a stressed cell or tissue. Proteomic based techniques that allow large-scale protein profiling are powerful tools for the identification of proteins involved in stress-responses in plants. Extensive studies have evaluated changes in protein levels in plant tissues in response to stresses. Unfortunately, these studies have been mainly focused on non-legume species such as *Arabidopsis* and rice, and only recently have been enlarged to include some legumes. As a result only a handful of studies have been carried out in legumes, although in the next few years there should be a significant increase in the number of legume species and stresses would be analyzed. Recently, proteomic approaches have been applied to various legumes like *M. truncatula*, lentils, lupin, common bean, cowpea and soybean to identify proteins involved in the response to different stresses. Interestingly, many of the induced proteins from these different stresses were common or belonged to overlapping pathways [13].

Considerable amount of research has been carried out during the last decade to find the effect of stress under extreme . These include chloroplast membrane, cell wall and nuclear envelope, while some researchers have focused on individual tissues *viz.* seeds, mitochondria, root tips, vacuoles, chloroplasts and thylakoids. To date, lots of reports have come which emphasize changes in protein expression levels during a particular or integrative stress consequently affecting cellular metabolism. Proteomics provides direct assessment of the biochemical processes of monitoring the actual proteins performing signaling, enzymatic, regulatory and structural functions encoded by the genome and transcriptome.

Following are the different categories of proteins with important properties, which have been shown to play a crucial role against abiotic environmental stress as well as biotic stress. The data so collected from various plants including soybean is based on 2-DE, mass spectrometry and bioinformatics tools.

(a) *Antioxidants Enzymes*

Reactive oxygen species (ROS) in plant cellulars are produced as a consequence of myriad stimuli ranging from abiotic and biotic stress, production of hormonal regulators, as well as cell processes such as polar growth and programmed cell death [14]. These reactive molecules are generated at a number of cellular sites, including mitochondria, chloroplasts, peroxisomes, and at the extracellular side of the plasma membrane. ROS trigger signal transduction events, such as mitogen-activated protein kinase cascades eliciting specific cellular response.s. The influence of these molecules on cellular processes is mediated by both the perpetuation of their production and their amelioration by scavenging enzymes such as superoxide dismutase, ascorbate peroxidase, and catalase. The location, amplitude, and duration of production of these molecules are determined by the specificity of the responses [15]. Accumulation of ROS as a result of various environmental stresses is a major cause of loss of crop productivity worldwide. ROS affect many cellular functions by damaging nucleic acids, oxidizing proteins, and causing lipid peroxidation. It is important to note that whether ROS will act as damaging, protective or signaling factors depends on the delicate equilibrium between ROS production and scavenging at the proper site and time. ROS can damage cells as well as initiate responses such as new gene expression. The cell response evoked is strongly dependent on several factors. The subcellular location for formation of ROS may be especially important for a highly reactive ROS, because it diffuses only a very short distance before reacting with a cellular molecule. Stress-induced ROS accumulation is counteracted by enzymatic antioxidant systems that include a variety of scavengers, such as superoxide dismutase, ascorbate peroxidase, glutathione peroxidase, glutathione S-transferase, catalase and non-enzymatic low molecular metabolites, such as ascorbate, glutathione (red.), α-tocopherol, carotenoids and flavonoids. In addition, proline can now be added to an elite list of non-enzymatic antioxidants that microbes, animals, and plants need to counteract the inhibitory effects of ROS [16]. Plant stress tolerance may therefore be improved by the enhancement of *in vivo* levels of antioxidant enzymes. The antioxidants as described are found in almost all cellular compartments which signify the importance of ROS detoxification for cellular survival. It has also been shown that ROS influence the expression of a number of genes and signal transduction pathways which suggest that cells have evolved strategies to use ROS as biological stimuli and signals that activate and control various genetic stress-response programs. Control of plant pathogens by genetic engineering has targeted ROS for development of pathogen resistant crop varieties [17]. Antisense technology has been used to reduce the capability to scavenge H_2O_2 in case of model plants like *Arabidopsis thaliana* and *Nicotiana tabacum*. In these plants, antioxidant enzymes like catalase and ascorbate peroxidase are under-expressed and it has been found that they were hyper-responsive to pathogen attack. This further confirms that the ability of plant cells to regulate the efficiency in their ROS-removal strategies is a key point in their resistance against pathogens. The technology is yet to be implemented in case of legumes including soybean as intensive research is going on future prospects of the technology.

(b) *Abscissic acid signaling and related protein*

Abscissic acid (ABA) has been implicated in plant response to environmental stress by interfering at different levels with signaling. Its level increases under stress conditions to trigger metabolic and physiological changes [18]. It has become increasingly clear that the isolated abiotic signaling network is controlled by ABA and the biotic network is controlled by salicylic acid, jasmonic acid and ethylene are interconnected at various levels [19]. The concept of marker genes whose expression is believed to be regulated by individual hormones does not do justice to the nature of the network. The apparent cross-talk in stress-hormone signaling makes it difficult to assign a marker gene or a mutant phenotype to a specific hormone-controlled pathway. The signaling network into which the four stress hormones and other signals feed is apparently designed to allow plants to adapt optimally to specific situations by integrating possibly conflicting information from environmental conditions, biotic stress, and developmental as well as nutritional status. Promoter analyses of ABA/stress-responsive genes revealed that a DNA sequence element consisting of ACGTGGC is important for ABA regulation. For the past several years, researchers have been trying to identify transcription factors that regulate the expression of ABA/stress-responsive genes *via* the consensus element, which is generally known as 'Abscisic Acid Response Element' (ABRE). Many basic leucine zipper class DNA-binding proteins that interact with the element have been reported [20]. Researchers have focused on the small subfamily of *Arabidopsis* basic leucine zipper proteins referred to as ABFs (ABRE-binding factors), whose expression is induced by ABA and by various abiotic stresses (i.e., cold, high salt and drought). ABA is involved in responses to environmental stress such as salinity, and is required by the plant for stress tolerance as found recently on soybean studies. The leaf ABA content in salt-tolerant soybean increased significantly under salt stress, while in case of salt sensitive soybean has almost negligible increase in ABA. It is thus possible that ABA enhances salt tolerance in soybean [21].

(c) *GABA-related protein*

γ-Aminobutyric acid (GABA) is a non-protein amino acid that is conserved from bacteria through yeast to vertebrates and was discovered in plants over half-a-century ago. It is mainly metabolized through a short pathway called the GABA shunt, because it bypasses two steps of the tricarboxylic-acid (TCA) cycle. The pathway is composed of three enzymes: the cytosolic and mitochondrial glutamate decarboxylase (GAD), GABA transaminase (GABA-T) and succinic semialdehyde dehydrogenase (SSADH). Although there are differences in the subcellular localization of GABA-shunt enzymes in different organisms have been reported (for e.g. in yeast, SSADH is present inside cytosol) [22]. In an alternative reaction, succinic semialdehyde can be converted to GHB (γ-hydroxybutyric acid) through a GHB dehydrogenase (GHBDH) present in animals and recently identified in plants [23]. Interestingly, research of GABA in vertebrates has focused mainly on its role in the context of plant responses to stress, because of its rapid and dramatic production in response to biotic and abiotic stresses. For example, disruption of the unique SSADH gene in *Arabidopsis* results in plants undergoing necrotic cell death caused by the accumulation of reactive oxygen intermediates (ROIs) when they are exposed to environmental stresses [24]. A recent article reports that a gradient of GABA concentration is essential for the growth and guidance of pollen tubes and suggests that this amino acid plays a role in intercellular signaling in plants, possibly similar to its role in animals. The main question raised by these recent findings is whether GABA itself serves as a signaling molecule in plants. If so, this would imply that GABA is capable of mediating developmental changes and cell guidance by interacting with specialized plant receptors [25].

(d) *Mitogen-activated protein kinase signaling and related proteins*

Like other eukaryotes, plants use mitogen-activated protein kinase (MAPK) cascades to regulate various cellular processes in response to a broad range of biotic and abiotic stress. These cascades promote the transient activation of MAPKs by a dual phosphorylation of Thr and Tyr within the activation loop of the MAPK. Recent studies indicate that MAPKs are not only regulated through phosphorylation by upstream kinases, but also by direct binding of different protein factors [26]. The constitutive activation of MAPKs was found to result in detrimental effects, underlining the importance of a negative regulation of MAPK signaling. MAPK phosphatases (MKPs) are negative regulators of MAPKs. Recent progress in analyzing plant MKP mutants has revealed their important role in fine-tuning MAPK signaling. In particular, the dual-specificity phosphatase MKP1 and the protein tyrosine phosphatase (PTP1) negatively regulate defense responses and resistance to a bacterial pathogen by counter balancing the activation of two MAPKs (MPK3 and MPK6). Interestingly, MKP1 and PTP1 bind CaM, and the phosphatase activity of MKP1 is increased by CaM in a Ca^{2+}-dependent manner. Thus, Ca^{2+} and MAPK signaling pathways appear to be connected through the regulation of plant MAPKs and MKPs by CaM [27].

(e) *Calcium signaling and related proteins*

Plant cells are equipped with highly efficient mechanisms to perceive, transduce and respond to a wide variety of internal and external signals during their growth and development. Perception of signals *via* receptors results in generation or synthesis of non-proteinaceous molecules which are termed as messengers. The messengers include Ca^{2+} ions, small organic molecules such as cyclic nucleotide monophosphates, inositol triphosphates and inorganic molecules such as H_2O_2 and NO. The elements of receptors, messengers, sensors and targets vary depending on the signal received. Identification and functional assignment of these elements in a stimulus-specific signal transduction pathway is a challenging area for plant biologists. With the completion of genome sequences of various organisms, including *Arabidopsis thaliana, Oryza sativa, Medicago trunculata, Glycine max* etc. it has become evident that plants have a large number of motifs containing helix-loop-helix which binds to Ca^{2+} [28]. Further, Ca^{2+} has been implicated in mediating various developmental processes (pollen tube growth, root-hair and lateral root development and nodulation), hormone regulated cellular activities (cell division and elongation, stomatal closure/opening), pathogen- and elicitor-induced defense related processes, and a variety of abiotic stress signal induced gene expression. However, the identity and functions of downstream transducers and mechanisms by which Ca^{2+} mediates a variety of cellular responses are just begin to unravel in plants. In plants, spatially and temporally distinct changes in cellular Ca^{2+} concentrations, designated as "Ca^{2+} signatures" that are evoked in response to different stimuli like drought, salt or osmotic stresses, temperature, light and plant hormones represent a central mechanistic principle to present defined stimulus-specific information [29]. These specific "Ca^{2+} signatures" are formed by the tightly regulated activities of channels and transporters at different membranes and cell organelles. While the identity and function of components of the Ca^{2+} extrusion system are rather well understood in plant cells, the molecular identity of Ca^{2+} specific influx channels has remained unknown. However, non-specific influx of Ca^{2+} mediated by ligand gated cation channels like cyclic nucleotide gated channels and glutamate receptor-like proteins contribute to different Ca^{2+} mediated cellular functions like the response to pathogens, pollen tube growth and abiotic stress. The unique structural composition of Ca^{2+} binding proteins and the complexity of the target proteins regulated by the Ca^{2+} sensors allow the plant to tightly control the appropriate adaptation to its ever changing environment. It is actually still not well understood about interface of information presentation by a specific Ca^{2+} signal and initiation of information decoding by Ca^{2+} sensors that represent a most critical step in specific information processing [30].

4. Significance of proteomics in soybean allergenicity

Soybeans have played a central role in concerns about GM introduced allergens and in using GM to remove intrinsic allergens. Soybean is a rich and inexpensive source of proteins for humans and animals. Soybean milk and dairy product replacement is growing in acceptance, not only by people sensitive to lactose and/or milk proteins, but also for health considerations. Soybean protein is widely used in thousands of processed foods throughout the industrialized world and is a staple crop in Asia. Soybean ranks among the eight most significant food allergens. Soybean sensitivity is estimated to occur in 5-8% of children and

1-2% of adults. The allergic reaction is only rarely life-threatening with the primary adverse reactions to consumption being atopic (skin) reactions and gastric distress. Symptoms of soy allergy usually appear within a few minutes to two hours of eating soy ingredients. People with soy allergies may cross-react with peanuts or other legumes, such as beans or peas. Soy is one of the most common allergens for infants who have not yet begun eating solid foods, because they may be fed soy-based infant formula. It is rare for babies to have a traditional IgE mediated food allergy to soy, but some babies may develop milk-soy protein intolerance [31-34] or food protein induced enterocolitis syndrome [http://foodallergies.about.com/od/soyallergies/a/Soy-Allergy-Overview.htm]. Infants will usually develop these sensitivities within a few months of birth, and most will outgrow them by the age of two. Most people with soy allergies can tolerate the small amount of soy protein that remains in refined soybean oil and soy lecithin. Both of these ingredients may cause allergic reactions in highly sensitized people. There are some data available that describe the natural variation in allergen proteins that occur in soybean. For a better understanding of the variation of allergen proteins that might be expected to occur in GM soybeans, it is important to determine the natural variation of protein composition both in wild and GM soybeans. "Proteomics" approach is the foremost one which allows protein identification and quantification with utmost accuracy.

Biotechnology critics have claimed that an apparent rise in the number of soybean allergic individuals in the UK is correlated with the development of GM soybeans in the American market. GM-soybeans that have been developed in the US include herbicide-resistance (glyphosate) and seeds with higher percentage of essential amino acids, *esp.* methionine. Experiments have directly tested the allergenicity of herbicide-tolerant soybeans using immunological tests with samples from soybean-sensitive people. These assays have shown that herbicide-resistant GM soybeans do not present any measurable differences in allergenicity compared with non-GM soybeans and are, therefore, substantially equivalent by allergenic criteria. Sensitive people remain allergic to GM soybeans, but there is no additional allergenic risk to others. According to some reports protein expressed corresponding to transgene responsible for herbicide-resistance in soybeans has allergenic motifs [35]. On ingestion a portion of the transgene along with the promoter get transferred to human gut bacteria. The transformed bacteria containing transgene continues to produce herbicide-resistance allergenic protein even when the individual is not eating GM soy. Therefore an individual is constantly exposed to potentially allergenic protein, being created within his gut. Further, herbicide-resistant protein is made more allergenic due to its misfolding brought by rearrangement of unstable transgenes. Some reports emphasize the fact that protein allergenicity is due to suppression of pancreatic-enzymes due to which protein remains in the gut for longer duration contributing to allergies. There is insufficient data to support *in vivo* toxicity of herbicide-resistant protein either due to transformation or enzyme suppression [36]. GM-soybeans with enhanced methionine content such as prolamines and 2S albumins were tested for its allergenicity before its commercialization. It was found that allergenicity was much higher with respect to wild soybeans [37]. Consequently the development of GM soybean with enhanced methionine has been abandoned and no product was released, thus nobody was harmed by its adverse reactions. Recently, one of the interesting analyses has been done on GM-soy irrespective of herbicide resistance or enhanced methionine content. It has found that GM transformation process may lead to increment in natural allergens in soybeans. The level of one known allergen is trypsin inhibitor which is 27% higher in raw GM soy varieties with respect to natural varieties [38]. Further, it has also been found that cooked GM soy has sevenfold higher amount of trypsin inhibitor as compared to cooked non-GM soy due to its extreme heat stability. There are several reports including both supportive as well unsupportive towards effects of GM-soy on humankind as well as on other flora and fauna of the environment. It will require intensive research including proteomics before their release into the commercial markets.

Plant biotechnology has not only tried to produce GM-soy which is herbicide resistance or with enhanced methionine content but also aimed to remove naturally occurring allergens in native soy varieties. Presently primary treatment for food allergies is avoidance, but it is unavoidable in case of soybean protein which is present in thousands of products. Therefore, it is very difficult to avoid soybean and its derived products. Research is going on to produce hypoallergenic variants of soybean which has potential to reduce the risk of adverse reactions. Soybeans possess as many as 15 proteins recognized by IgEs from soybean-sensitive people [39]. The immunodominant soybean allergens are the β-subunit of conglycinin and P34 or Gly m Bd 30k (cysteine proteases from papain family). The P34/Gly m Bd 30k protein is a unique member of the papain superfamily lacking the catalytic cysteine residue that is replaced by a glycine which is 70% more allergenic with respect to conglycinin. There are several approaches that have been taken to produce a hypoallergenic soybean. One approach was to search cultivars which lack allergens and then crossing its germplasm to elite germplasm. This approach could not be implemented as there was no soybean cultivar (either domesticated or wild) present which lack P34/Gly m Bd 30k. Immunological assays of P34/Gly m Bd 30k with antibodies from soybean-sensitive people resulted in the identification of 14 contiguous and non-contiguous linear epitopes. The presence of so many distinct linear epitopes means that the probability of a naturally occurring variant with a sufficient number of alterations to disrupt the allergenicity is extremely small. Protein engineering could be performed to alter amino acid sequence by disrupting allergenic sequences. Using linear peptides to test possible modifications, it is straightforward to assay numerous variants and pick one that is not recognized by the IgE population. The epitope modification approach is not feasible to produce an essentially hypoallergenic variant. The problem with this technology is to remove completely the intrinsic allergen and substitute the 'hypoallergenic' variant in its place. Further, the modification of the protein to remove the allergenic epitopes may alter the protein's folding, that, in turn, may affect the protein's intracellular targeting, stability and accumulation. All these possibilities will need to be tested for experimentally and, finally; the newly produced hypoallergenic variant will need to be tested to ensure that it too is not a new allergen. For these reasons, substituting a hypoallergenic variant of a plant still has a high technological threshold and has yet to be achieved. The alternative GM approach is to eliminate the allergen by suppression. There have been several attempts to reduce and/or eliminate allergens using gene suppression technology. Gene-silencing techniques involve transgenic soybeans with eliminated immunodominant human allergen P34/Gly m Bd 30k. It involves complete elimination of the P34/Gly m Bd 30k allergen from the initial somatic embryos through the third generation homozygous soy-beans. Suppression of the allergen did not introduce any changes in the pattern of growth and development of the plant or seed at both the gross and subcellular level. In order to compare the P34-suppressed soybeans with the wild type, large-scale proteomic analysis was performed. Imaging of the 2D gels identified over 1400 individual elements. Mass spectrometry analysis of about 140 of these spots confirmed that the only overt changes in composition in the transgenic soybeans was the suppression of the P34/Gly m Bd 30k protein with no other proteins induced or suppressed [40]. Further analysis with sera samples from soybean-sensitive people confirmed a loss of the P34 allergen and no induction of any new allergens. The proteome and immunological analysis together confirms that it is feasible to suppress an endogenous allergen without introducing adverse effects on the plant or changing the composition of the soybean seed in any way other than the removal of the targeted protein. This result meets the test of `substantial equivalence' where the GM soybean seed is essentially identical except for the change in the single desired characteristic. Suppressing P34/Gly m Bd 30k in GM soybeans is a first step and a demonstration in addressing the growing concerns about food allergies and its relationship to the development of GM crops. More detailed studies and approaches should provide the tests needed to gain regulatory approval in nations that are currently cautious about this technology. Natarajan *et al.* [41] have compared the profiles of allergen and anti-nutritional proteins both in wild and GM soybean seeds. 2D-PAGE was used for the separation of proteins at two different pH ranges and applied a combined MALDI-TOF-MS and LC-MS analysis for the identification of proteins. Although overall distribution patterns of the allergen and anti-nutritional proteins Gly m Bd 60K (conglycinin), Gly m Bd 30K, Gly m Bd 28K, trypsin inhibitors, and lectin appeared similar, there was remarkable variation in the number and intensity of the protein spots between wild and GM soybean. The wild soybean showed fifteen polypeptides of Gly m Bd 60K and three polypeptides of trypsin inhibitors. GM soybean showed twelve polypeptides of Gly m Bd 60K and two polypeptides of trypsin inhibitors. In contrast, the GM soybean showed two polypeptides of Gly m Bd 30K and three polypeptides of lectin and the wild type showed two and one polypeptides of Gly m Bd 30K and lectin, respectively. The same number of Gly m Bd 28K spots was observed in both wild and GM soybean [41].

The fear of allergic reactions has produced much of the concern about the risks of GM crops. In order to broadly apply genetic modification to crops, there is an urgent need for better biochemical and molecular methods, including animal models, to test for food allergens experimentally so that the supporting data can be provided to evaluate newly proposed and actual GM products. In order to design transgenes, it would be useful to predict allergenicity but, currently, there are no models that would permit accurate assessment of allergenic potential of proteins unrelated to known allergens. Liver represents a suitable model for monitoring the effects of a diet, due to its key role in controlling the whole metabolism. Previous studies on hepatocytes from young female mice fed on GM soybean demonstrated nuclear modifications involving transcription and splicing pathways [42, 43]. The morpho-functional characteristics of the liver of 24-month-old mice, fed from weaning on control or GM soybean, were investigated by combining a proteomic approach with ultrastructural, morphometrical and immunoelectron microscopical analyses. Several proteins belonging to hepatocyte metabolism, stress response, calcium signaling and mitochondria were differentially expressed in GM-fed mice, indicating a more marked expression of senescence markers in comparison to controls. Moreover, hepatocytes of GM-fed mice showed mitochondrial and nuclear modifications indicative of reduced metabolic rate. This study demonstrates that GM soybean intake can influence some liver features, although the mechanisms remain unknown. Therefore, it is required to investigate the long-term consequences of GM-diets, further studies are required for potential synergistic effects with other factors like ageing, stress etc.

5. Challenges and perspectives

Soybean is a species of great agronomic and economic interest. It is one of the most recalcitrant plant species to be used as experimental material in proteomic analysis. Furthermore, there are several difficulties in the study of proteins (irrespective of source) with respect to DNA and RNA. The foremost important thing is the maintenance of secondary and tertiary structure during their analysis. They have problems with easy denaturation on exposure to high temperature, extremes of pH, oxidation, specific chemicals etc. There are some classes of proteins which are difficult to analyze due to their poor solubility. Proteins cannot be amplified like DNA, therefore less abundant species are very difficult to detect. However, many potentially important proteins (in scarce) are lost due to non-specific binding or the co-removal of proteins/peptides intrinsically bound to the high abundant carrier proteins. Following are two methods developed recently to resolve detection of less abundant plant proteins [44]:

- The use of equalizer beads coupled with a combinational library of ligands containing diverse population of beads with equivalent binding capacity to most of the proteins present in a sample.

- The ultra-microarrays have been found to have high specificity and sensitivity with detection levels in the range of attomole (10^{-18} mole).

The current depth of knowledge regarding the soybean proteome is significantly less than that for some other plants. The soybean proteome map which is available in the database (http://proteome.dc.affrc.go.jp/soybean/) corresponds to various types of stresses, allergenicity, and studies on natural product biosynthesis in soybean. The other challenges in plant proteomics including soybean are standardization of methodologies, dissemination of proteomics data into publicly available databases and most importantly its cost expensiveness. Furthermore, most proteomics technologies use complex instrumentation and critical computing power. Currently, there is no expertise available for functional interpretation of data obtained from integration of proteomics with genomics and metabolomics.

The significance of proteomics over genomics and transcriptomics has been debated since the field has emerged. The importance of the proteome cannot be overstated as it is the proteins within the cell that provide structure, produce energy, as well as allow communication, movement, and reproduction. Basically, proteins provide structural and functional framework for cellular life. Genetic information is static while the protein complement of a cell is dynamic. Differential proteomics is a scientific discipline that detects the proteins associated with a diseased state (either due to abiotic or biotic stress, toxicity due to allergenicity, genetic modifications etc.) by means of their altered levels of expression between the control and diseased states. Extensive research towards the development of a soybean proteome map would permit the rapid comparison of soybean cultivars, mutants, and transgenic lines. Moreover, studies of soybean physiology will also benefit from the existence of a detailed and quantitative proteome reference map of the soybean plant. The information obtained from soybean proteomics will be helpful in predicting the function of plant proteins and will aid in molecular cloning of the corresponding genes in the future. The identification of novel genes, the determination of their expression patterns in response to stress, and an understanding of their functions in stress adaptation will provide us with the basis for effective strategies for engineering improved stress tolerance in soybean. With the advancement of new technologies in proteomics combined with advanced bioinformatics, we are currently identifying molecular signatures of diseases based on protein pathways and signaling cascades. Applying these findings will improve our understanding of the roles of individual proteins or the entire cellular pathways in the initiation and development of disease. The abundance of information provided by proteomics research is entirely complementary with the genetic information being generated by genomics research. Proteomics makes a key contribution to the development of functional genomics. The combination of genomics and proteomics will play a major role in understanding molecular mechanisms in plant pathology, and it will have a significant impact on the development of high yield varieties, with better resistance towards adverse environmental factors as well as various pathogenic diseases caused by bacteria, viruses and fungi in the future.

Author details

Alka Dwevedi[1*] and Arvind M Kayastha[2*]

*Address all correspondence to: kayasthabhu@gmail.com

1 Regional Centre for Biotechnology, India

2 School of Biotechnology, Faculty of Science, Banaras Hindu University, India

References

[1] Wu, D. D., Hu, X., Park, E., Wang, X., Feng, J., & Wu, X. (2010). Exploratory analysis of protein translation regulatory networks using hierarchical random graphs. *BMC Bioinformatics*, 11(3), S2.

[2] Ezkurdia, I., del Pozo, A., Frankish, A., Rodriguez, J. M., Harrow, J., Ashman, K., Valencia, A., & Tress, M. L. (2012). Comparative proteomics reveals a significant bias toward alternative protein isoforms with conserved structure and function. *Molecular Biology and Evolution*, 29-2265.

[3] Bertone, P., & Snyder, M. (2005). Prospects and challenges in proteomics. *Plant Physiology*, 138-560.

[4] Hefle, S. L, Nordlee, J. A, & Taylor, S. L. (1996). Allergenic foods. *Critical Reviews in Food Science and Nutrition*, 36, S69-S89.

[5] Metcalfe, D. D. (2005). Genetically modified crops and allergenicity. *Nature Immunology*, 6-857.

[6] Manavalan, L. P, Guttikonda, S. K, Tran, L. S, & Nguyen, H. T. (2009). Physiological and molecular approaches to improve drought resistance in soybean. Plant & Cell Physiology , 50-1260.

[7] Atkinson, N. J, & Urwin, P. E. (2012). The interaction of plant biotic and abiotic stresses: from genes to the field. *Journal of Experimental Botany*, 63-3523.

[8] Nouri-Z, M., Toorchi, M., & Komatsu, S. (2011). Chapter 9: Proteomics approach for identifying abiotic stress responsive proteins in soybean. *In Soybean-molecular aspects of breeding ed. Aleksandra Sudaric. Intech, April 11*, 978-9-53307-240-1.

[9] Mooney, B. P., Krishnan, H. B., & Thelen, J. J. (2004). High-throughput peptide mass fingerprinting of soybean seed proteins: automated workflow and utility of UniGene expressed sequence tag databases for protein identification. *Phytochemistry*, 65-1733.

[10] Natarajan, S., Xu, C., Caperna, T. J., & Garrett, W. M. (2005). Comparison of protein solubilization methods suitable for proteomic analysis of soybean seed proteins. *Analytical Biochemistry*, 342-214.

[11] Hajduch, M., Ganapathy, A., Stein, J. W., & Thelen, J. J. (2005). A systematic proteomic study of seed filling in soybean. Establishment of high-resolution two dimensional reference maps, expression profiles, and an interactive proteome database. Plant Physiology , 137-1397.

[12] Lei, Z, Elmer, A. M., Watson, B. S., Dixon, R. A, Mendes, P. J., & Sumner, L. W. (2005). A two-dimensional electrophoresis proteomic reference map and systematic identification of 1367 protein from a cell suspension culture of the model legume Medicago truncatula. *Molecular and Cellular Proteomics*, 4, 1812-1825.

[13] Nasi, A., Picariello, G., & Ferranti, P. (2009). Proteomic approaches to study structure, functions and toxicity of legume seeds lectins. Perspectives for the assessment of food quality and safety. *Journal of Proteomics*, 72-527.

[14] Bailey-Serres, J., & Mittler, R. (2006). The roles of reactive oxygen species in plant cells. *Plant Physiology*, 141, 311.

[15] Mittler, R. (2002). Oxidative stress, antioxidants and stress tolerance. *Trends in Plant Science*, 7-405.

[16] Daft, J., Vandenabeele, S., Vranová, E., Van Montagu, M., Inzé, D., & Van Breusegem, F. (2000). Dual action of the active oxygen species during plant stress responses. *Cellular and Molecular Life Sciences*, 57-779.

[17] Collinge, D. B., Jørgensen, H. J .L., Lund, O. S., & Lyngkjær, M. F. (2010). Engineering pathogen resistance in crop plants: Current Trends and Future Prospects. *Annual Review of Phytopathology*, 48-269.

[18] Leung, J., & Giraudat, J. (1998). Abscisic acid signal transduction. *Annual Review of Plant Physiology and Plant Molecular Biology*, 49-199.

[19] Mauch-Mani, B., & Mauch, F. (2005). The role of abscisic acid in plant-pathogen interactions. *Current Opinion in Plant Biology*, 8-409.

[20] Choi, H. I., Hong, J. H., Ha, J. O., Kang, J. Y., & Kim, S. Y. (2000). ABFs, a family of ABA-responsive element binding factors. *The Journal of Biological Chemistry*, 275-1723.

[21] Xu , Y. X., Fan, R., Zheng, R., Li, M. C., & Yu, Y. D. (2011). Proteomic analysis of seed germination under salt stress in soybeans. *Journal of Zhejiang University-Science B (Biomedicine & Biotechnology)*, 12-507.

[22] Bouche', N., Lacombe, B., & Fromm, H. (2003). GABA signaling: a conserved and ubiquitous mechanism. *Trends in Cell Biology*, 13-607.

[23] Roberts, M. R. (2007). Does GABA Act as a Signal in Plants? Plant Signaling & Behavior , 2-408.

[24] Bouché, N., Fait, A., Bouchez, D., Møller, S. G., & Fromm, H. (2003). Mitochondrial succinic-semialdehyde dehydrogenase of the γ-aminobutyrate shunt is required to restrict levels of reactive oxygen intermediates in plants. *Proceedings of National Academy of Sciences of United States of America*, 100-6843.

[25] Ludewig, F., Hüser, A., Fromm, H., Beauclair, L., & Bouché, N. (2008). Mutants of GABA transaminase (POP2) suppress the severe phenotype of succinic semialdehyde dehydrogenase (SSADH) mutants in Arabidopsis. Plos One , 3, e3383.

[26] Wrzaczek, M., & Hirt, H. (2001). Plant MAP kinase pathways: how many and what for? *Biology of the Cell*, 93-81.

[27] Zhang, T., Liu, Y., Yang, T., Zhang, L., Xu, L., & An, L. (2006). Diverse signals converge at MAPK cascades in plants. *Plant Physiology and Biochemistry*, 44-274.

[28] Yang, T., & Poovaiah, B. W. (2003). Calcium/calmodulin-mediated signal network in plants. *Trends in Plant Science*, 8-505.

[29] Reddy, V. S., & Reddy, A. S. N. (2004). Proteomics of calcium-signaling components in plants. *Phytochemistry*, 65-1745.

[30] Perochon, A, Aldon, D, Galaud, J. P., & Ranty, B. (2011). Calmodulin and calmodulin-like proteins in plant calcium signaling. *Biochimie*, 93, 2048-2053.

[31] Dwevedi, A., & Kayastha, A. M. (2010). Plant β-galactosidases: Physiological significance and recent advances in technological applications. *Journal of Plant Biochemistry and Biotechnology*, 19-9.

[32] Dwevedi, A., & Kayastha, A. M. (2011). Chapter 11: Soybean, a multifaceted legume with enormous economic capability. *In Soybean: Biochemistry, Chemistry and Physiology, ed. Tzi-Bun Ng. Intech*, April 26, 978-9-53307-219-7.

[33] Dwevedi, A., & Kayastha, A. M. (2009). A β-galactosidase from pea seeds (PsBGAL): purification, stabilization, catalytic energetics, conformational heterogeneity, and its significance. *Journal of Agricultural and Food Chemistry*, 57-7086.

[34] Kishore, D., Talat, M., Srivastava, O. N., & Kayastha, A. M. (2012). Immobilization of β-galactosidase onto functionalized graphene nano-sheets using response surface methodology and its analytical applications towards milk and whey lactose. *Plos One*, 7, e40708.

[35] Burks, A.W., & Fuchs, R. L. (1995). Assessment of the endogenous allergens in glyphosate tolerant and commercial soybean varieties. *Journal of Allergy and Clinical Immunology*, 96 1008.

[36] Mazur, B. J, & Falco, S. C. (1989). The development of herbicide resistant crops. *Annual Review of Plant Physiology and Plant Molecular Biology*, 40-441.

[37] Herman, E. M. (2003). Genetically modified soybeans and food allergies. *Journal of Experimental Botany*, 54-1317.

[38] Domingo, J. L. (2007). Toxicity studies of genetically modified plants. *Critical Reviews in Food Science and Nutrition*, 47721-733.

[39] Burks, A. W., Brooks, J. R, & Sampson, H. A. (1988). Allergenicity of major component proteins of soybean determined by enzyme-linked immunosorbent assay (ELISA) and immunoblotting in children with atopic dermatitis and positive soy challenges. *Journal of Allergy and Clinical Immunology*, 81-1135.

[40] Helm, R. M., Cockrell, G., Herman, E., Burks, A. W., Sampson, H. A., & Bannon, G. A. (1998). Cellular and molecular characterization of a major soybean allergen. *International Archive of Allergy Immunology*, 117-29.

[41] Natarajan, S. S., Xu, C., Bae, H., Caperna, T. J., & Garrett, W. M. (2006). Proteomic analysis of allergen and antinutritional proteins in wild and cultivated soybean. *Journal of Plant Biochemistry and Biotechnology*, 15-103.

[42] Malatesta, M., Caporaloni, C., Gavaudan, S., Rocchi, M. B., Serafini, S., Tiberi, C., & Gazzanelli, G. (2002). Ultrastructural morphometrical and immunocytochemical analyses of hepatocyte nuclei from mice fed on genetically modified soybean. *Cell Structure and Function*, 27-173.

[43] Malatesta, M., Boraldi, F., Annovi, G., Baldelli, B., Battistelli, S., Biggiogera, M., & Quaglino, D. (2008). A long-term study on female mice fed on a genetically modified soybean: effects on liver ageing. *Histochemistry and Cell Biology*, 130-967.

[44] Cho, W. C. S. (2007). Proteomics technologies and challenges. *Genomics, Proteomics & Bioinformatics*, 5-77.

Chapter 3

Advancements in Transgenic Soy: From Field to Bedside

Laura C. Hudson, Kevin C. Lambirth,
Kenneth L. Bost and Kenneth J. Piller

Additional information is available at the end of the chapter

1. Introduction

Today biotechnology and the process of genetic modification is emerging and advancing worldwide. At the forefront of this technology is soybean, which has become a popular subject of genetic modification due to its versatility and economical importance as a crop plant. Over 15 years ago the first herbicide resistant soybeans were introduced into the market. By 1997, approximately 8% of all soybeans cultivated for commercial use in the United States were genetically modified. This trend has grown exponentially and by 2011 the percentage of genetically modified soybean rose to 94% in the United States and 81% worldwide. The technology to genetically modify soybean has not only had a huge impact on the commercial agricultural market, but has paved the way to an onset of both traditional and nontraditional uses for soybean as well as opening up many new potential applications for this important crop plant. Soybean has become a popular subject of genetic modification over the past two decades and with the advancement of plant transformation technology, it is now possible to manipulate and or add various traits to soybean.

2. Soybean transformation

There are several protocols used to genetically transform plants with either stable or transient expression. Some of the methods include electroporation, silicon carbide fibers, liposome mediated transformation and *in planta Agrobacterium*-mediated transformation via vacuum infiltration of whole plant. However, many of these methods are not used for soybean transformation because of low transformation efficiencies. Two more commonly used platforms that have been successfully optimized for stable soybean transformation include cotyledonary node–*Agrobacterium*-mediated transformation and somatic embryo–particle-bombardment-mediated transformation.

The *Agrobacterium*-mediated plant transformation method uses a soil dwelling bacteria species called *Agrobacterium tumefaciens* to transfer desirable genes into plants. Using this method, a foreign gene can be placed within the T- DNA boarder regions of the bacterial plasmid which then integrates into a host plant's genome [1]. Wounded plant tissue gives off specific phenolic compounds which induce *Agrobacterium* to express a set of virulence (vir) genes. The expression of the vir genes results in the production of single-stranded DNA that is transferred and integrated into the plant genome.

There are several advantages of *Agrobacterium*-meditated plant transformation including straight forward methodology, minimal equipment cost, and reliable insertion of a single or a low copy transgene number. The first reported transformation of soybean with an *Agrobacterium* strain used co-cultivation followed by organogenesis from cotyledonary nodes [2]. This work was followed by using *Agrobacterium* mediated transformation of immature cotyledons [3], and embryogenic suspension cultures [4]. Since then, several groups have worked to improve these methods, in particular the transformation and regeneration from cotyledonary nodes. Cotyledonary node regions contain axillary meristems at the junction between cotyledon and hypocotyl. Generally, the cotyledonary nodes are pre-wounded and then co-cultivated with *Agrobacterium*. The axillary meristems proliferate and regenerate through the formation of multiple adventitious shoots on culture medium containing a cytokinin. In the United States, public facilities, including the Plant Transformation Facility at Iowa State University and the Plant Transformation Core Facility at the University of Missouri, provide fee for service genetic transformation of soybean for public research, mainly by cotyledonary node *Agrobacterium*-mediated transformation.

The other widely used method of soybean transformation is somatic embryo particle bombardment-mediated transformation also called particle bombardment, or biolistic technology. This method directs small tungsten or gold particles coated with the desired genes toward the target plant cells with enough force to penetrate the cell wall and membrane [5]. Once inside the cell the DNA disassociates from the particle and becomes integrated into the plant genome.

The particle bombardment transformation method was first used in soybean in 1988 by McCabe et al., who successfully transformed immature seed meristem [6] and was followed by the transformation of somatic embryonic tissue [7], and apical meristem [8]. Transformation of somatic embryos has been the most successful method and is induced from immature cotyledons cultured on medium containing moderately high concentrations of an auxin. These cotyledons are used to generate proliferative embryogenic cultures and to recover whole plants. A major advantage of the particle bombardment transformation method relative to *Agrobacterium*-based methods is the removal of biological incompatibilities between tissues of many plant species and the *Agrobacterium* vector. However, it has been shown that certain genotypes are more susceptible to the formation of proliferative embryogenic tissue than others. Limitations of the bombardment process include the requirement of specialized equipment (gene gun), transformation limited to cells at or near the surface, and high copy number events with high levels of recombination which may not be desirable.

The development of soybean transformation methods has paved the way for an extensive amount of research to develop genetically modified soybeans that have been widely adopted for crop improvement purposes. This has been a fast growing field with the addition of many agronomic, nutraceutical, and pharmaceutical traits being developed, the progress of which will be reviewed in this chapter.

3. Agronomical improvements in soybean

3.1. Herbicide tolerance

In 1970, glyphosate, a broad-spectrum foliar herbicide, was discovered [9]. Glyphosate inhibits 5-enolpyruvylshikimate-3-phosphate synthase (EPSPS), a crucial enzyme of the shikimate biosynthetic pathway that is responsible for the production of several essential aromatic amino acids. Glyphosate was commercially introduced in 1974 and by 1995 use had reached 4. 5 million kg in the United States. Due to popularity and broad use of this herbicide by farmers, glyphosate became a candidate for research in creating herbicide resistant soybeans and has led to one of the most well-known examples of herbicide tolerance: The Roundup Ready® soybean developed by Monsanto. Roundup Ready® soybeans were one of the first examples of a commercially viable transgenic plant. These transgenic soybeans express functional EPSPS providing tolerance to the herbicide glyphosate (Roundup™). The popularity of these soybeans grew with farmers since Roundup™ could be applied to a field of Roundup Ready® soybeans to significantly reduce weed populations while leaving the soybean crop unharmed. The development of Roundup Ready® soybeans offered farmers many advantages in a system that was relatively easy to use. The level and consistency of weed control allowed farmers to take advantage of a no-till system, and eliminated the need for cultivation allowing growers to space rows more closely. Narrow row planting results in higher yields due to a more efficient use of space and may result in better weed control, as the canopy closes more quickly providing earlier competition against weeds. In addition, the window of application for Roundup™ is wider than for other post emergence herbicides currently used in soybeans, both in terms of the stage of soybean growth and the ability to achieve effective control of larger weeds. These factors contributed to the popularity of this weed control technology. At present, approximately 90% of the soybeans farmed in the United States utilize this technology.

In response to Monsanto's hugely successful Roundup Ready® crops, Bayer Crop Science released its own herbicide tolerant soybean known as Liberty Link® soybean [10]. Liberty Link® soybeans were developed to express a gene derived from the bacteria *Streptomyces viridochromogenes* called phosphinothricin-N-acetyltransferase (PAT). PAT is a glutamine synthetase inhibitor that binds to glutamate, making plants resistant to the broad-spectrum contact herbicide glufosinate ammonium. This herbicide causes cessation of photosynthesis and plant death by interfering with the biosynthetic pathway of the amino acid glutamine and with ammonia detoxification. Glufosinate ammonium is the active ingredient in phosphinothricin herbicides (Basta®, Ignite®, Rely®, Liberty®, Harvest®, and Finale®) used to control a wide range of weeds after the crop emerges or for total vegetation control on land not used for cultivation. Since glufosinate ammonium-based herbicides function by a different mode of action than glyphosate-based herbicides, the Liberty Link® system provides farmers with an alternative strategy for controlling weeds.

There are several other examples of transgenic soybeans expressing herbicide resistance traits that are in various stages of development. Pioneer has developed a transgenic soybean product that provides tolerance to two different classes of herbicides: glyphosate and acetolactate synthase (ALS)-inhibiting herbicides. These soybean plants express the glyphosate acetyltransferase (GAT4601) and modified version of a soybean acetolactate synthase (GM-HRA) proteins. [11]. The GAT4601 protein confers tolerance to glyphosate-containing herbicides by acetylating glyphosate and thereby rendering it non-phytotoxic. The GM-HRA protein confers tolerance to the ALS-inhibiting class of herbicides. The development of GM soybeans with characteristics controlled by multiple genes leading to the expression of two herbicides is a different approach than previous strategies involving single characteristics controlled by a single gene. These genetically modified soybeans express a combination of herbicides with different modes of action. Inherent crop tolerance will enable more effective management of weed populations.

BASF has used a similar method to introduce a soybean that combines herbicide-tolerant soybean varieties with the broad spectrum imidazolinone class of herbicides. These transgenic soybeans contain the csr1-2 gene derived from *Arabidopsis thaliana* that encodes the imidazolinone-tolerant AHAS-Large subunit (also known as ALS). The AHAS-L subunit interacts with the endogenous soybean small regulatory subunit to form an enzyme complex that catalyzes the first step in the synthesis of branched-chain essential amino acids, valine, leucine, and isoleucine [12]. The AHAS enzymes occur ubiquitously in plants. Imidazolinone herbicides inhibit the native enzymes, resulting in plant death.

To address the potential emergence of other herbicide resistant broadleaf weeds, Monsanto has developed a line of transgenic soybeans that are resistant to treatment with dicamba [13]. Dicamba (3,6-dichloro-2-methoxybenzoic acid) is a low-cost, widely-used, broad leaf herbicide that is environmentally friendly. Soybeans transformed with a genetically engineered bacterial dicambamonooxygenase (DMO) gene were able to inactivate dicamba, making them resistant to this herbicide. Dicamba-resistant soybeans are in the advanced stages of research and development and are predicted to be commercialized soon.

Syngenta and Bayer CropScience are co-developing HPPD-inhibitor tolerant soybeans, a novel herbicide tolerance trait for soy. The event consists of a molecular stack of a gene conferring tolerance to hydroxyphenylpyruvatedioxygenase (HPPD)-inhibiting herbicides as well as a gene for glufosinate tolerance. Inhibition of HPPD stops the catabolic degradation of tyrosine to plastoquinones which is important for photosynthesis, carotenoid biosynthesis, and tocopherol production [14]. This multiple herbicide tolerance stack will enable the use of multiple herbicides and will be an important new tool for soybean growers faced with increasing pressure from resistant weeds. In the future, other innovative molecular strategies can be expected to generate genetically modified (GM) soybeans with novel features to combat weeds and enhance weed resistance. These new GM soybeans will reduce environmental contamination risks and reduce costs for consumers and producers.

3.2. Insect resistance

Insect pest management through the use of chemicals has brought about considerable protection to crop yields over the past several decades. Unfortunately, extensive and indiscriminate usage of chemical pesticides has resulted in environmental degradation, adverse effects on human health and other organisms, eradication of beneficial insects, and development of pest-resistant insects. As farmers move forward with the objective of achieving greater crop productivity it will be imperative to replace chemical inputs with safer alternatives to manage insect pests in agricultural ecosystems. Within agricultural biotechnology, insect resistance is a prime research area that has potential to improve agricultural productivity and provide much needed alternatives to pesticides while being effective against pests, innocuous to non-target organisms, and cost effective. With the advent of biotechnology, the ability to genetically modify plants for insect resistance on a commercial scale is within reach. One of the most extensively studied traits for insect resistance in soybeans involves the Bt gene.

Bacillus thuringiensis (Bt) is a common bacteria found in the environment. It has been used as a biological control agent against lepidopteran insects for more than 50 years. Bt targets a class of compounds responsible for insecticidal activity known as crystalline proteins, or cry proteins (Cry1), that are highly toxic after ingestion. The mode of action for Cry1 toxins is the disruption of midgut cellular membranes leading to cell death. One of the primary advantages of using Bt genes for insect control in transgenic plants is the specific insecticidal action toward insects from the Lepidoptera order leaving beneficial insects, birds, and mammals unharmed. Thus, the insertion of Bt toxins into plants, by genetic modification, is an attractive model for the creation of insect resistant transgenic crops.

To date, many different plant species have been genetically modified to exhibit insect resistance using Bt. While the Bt trait has been commercialized in corn and cotton, it is still in developmental stages in soybean. Transformation of soybean with Bt to induce resistance to lepidopteron species has been performed for over a decade. By 1994 fertile transformed soybeans containing a synthetic Bt (Cry1Ac) were generated [15, 16]. Stewart et al., used detached leaf bioassays to show that transgenic soybean lines were resistant to multiple soybean pests with less than 3% leaf defoliation compared to 20% observed in traditionally bred lepidopteron resistance soybean lines [16].

Other groups have used a similar strategy by evaluating soybeans engineered with Cry1Ac for resistance to lepidopteron species under field conditions. One example compared Bt lines to controls in the field using field cages and artificial infestation with lepidopteron larvae over a three year period [17]. In this case, Bt lines showed up to 9 times less defoliation from pets when compared to control plants. Similarly, Mcpherson and MacRae reported the evaluation of Bt soybean lines for suppression of lepidopteron species in the field over 2 years [18]. In this case, soybean plants expressing Cry1Ac were essentially absent of lepidopteron populations when compared to peak population densities of 20-30 larvae per row in control plots. Furthermore, Bt lines showed <1.5% defoliation when compared to 53% defoliation in control plants.

The utility for Bt soybeans has become evident. This has lead to pyramiding strategies using Cry1Ac with native plant resistance genes to increase plant resistance against insect-pests. Several quantitative trait loci (QTLs) from soybean lines have been described as showing antixenosis and antibiosis resistance towards lepidopteron insects [19, 20]. This work lead to the development of transgenic soybean lines by combining QTLs with synthetic Cry1Ac [21]. In this case, field evaluations and detached leaf bioassays were used to test this multiple resistance gene pyramiding strategy for antibiosis resistance. Based on defoliation in the field, as well as larval weight gain on detached leaves, soybean lines carrying a combination of Cry1Ac and the QTL were significantly more resistant to lepidopteron pests.

While Bt soybean varieties have not been commercialized this body of research has lead Monsanto to the development of soybeans that incorporate the Bt trait stacked with the second generation Roundup Ready germplasm [22]. Bt Roundup Ready 2 Yield seeds are currently in Phase IV trials and are targeted for commercialization in Brazil in 2013. This pyramiding strategy would be the first in-seed insect protection for soybeans and is expected to offer an important technology for farmers who face significant yield loss due to insect damage. Although not universal in its application and total in its protection, Bt will play a central role in protecting the crop from major insect pests.

With the onset and success of Bt crops other avenues have been explored for their possible roles in the development of transgenic insect resistant plants. These approaches include the use of plant defense proteins, lectins, α-amylase inhibitors, insect chitinases, and defensins. The development and implementation of engineered insecticidal soybean varieties is currently in its infancy. The incorporation of a multiple gene stacking strategy will also be important in the future development of insect resistant soybeans. Bt, in combination with other biopesticides, has the potential to drastically reduce the consumption of chemical pesticides, however it will be important to continue research and have a development strategy for a future generation of technology, to ensure that insects do not rapidly develop resistance.

3.3. Disease resistance

The United States, Brazil, and Argentina are the three major soybean-producing countries in the world where more than 50% of all soybeans are harvested. Such a geographic distribution facilitates the spread of insect-pests and diseases. Hence, soybean can be attacked by many different pathogens, including bacteria, viruses, fungi, and nematodes. These pathogens and pests can cause damage in seeds, roots, leaves, stems and pods, and usually are tissue-specific. Therefore, disease resistance is another area of great interest for both researchers and farmers.

Disease control management is currently concentrated on agronomic practices, like planting under tillage, use of lodging resistant varieties, wide row planting, and rotation with non-host crops. Chemical control has poor efficiency because of low penetration and uneven distribution due to an already formed canopy. In addition, chemical application can be extremely expensive for farmers and unhealthy for the environment. There has been little success with conventional plant breeding for disease resistance in soybean leaving room for other approaches such as the use of biotechnology to produce genetically modified soybeans that have disease resistance.

3.3.1. Bacterial

Bacterial infections are widespread diseases that occur mainly in the mid-to-upper and young leaves of the soybean plant. There are several bacteria which cause disease in soybean resulting in large amounts of yield loss and poor seed quality. While there has been promising research in the development of bacterial disease resistance with the use of biotechnology in other crop plants such as rice, tomato, banana, and tobacco, there has been less research on the development of bacterial disease resistance for soybean. This research may lead to new strategies for the development of bacterial disease resistance in soybean.

3.3.2. Viral

The development of transgenic soybean that confers viral resistance has been studied a bit more extensively. Viruses in soybean are global pests. Significant resistance to several viruses in a number of plant species have been achieved through pathogen derived resistance by the use of viral coat proteins which, when expressed *in planta*, can interfere with viral assembly. This is the same approach that has been used by several groups to develop transgenic viral resistance in soybean. One of the first groups to investigate this approach was Di et al., who produced a soybean that was resistant to bean pod mottle virus (BPMV) [23]. This was done by introducing a BPMV coat protein into the soybean genome. Transgenic events showed complete resistance to BPMV infection. Another study created soybean lines that were resistant to BPMV by inserting a BPMV capsid polyprotien. Events generated in this case were subjected to infectivity assays and not only exhibited resistance to virus infection, but also exhibited systemic infection, showing little to no visible symptoms [24]. Transgenic lines such as these could lead to future commercial cultivars with resistance to BPMV.

The development of soybean mosaic virus (SMV) resistant soybeans is important since SMV is found in all regions where soybean is grown and infection can cause yield loss up to 90%. Despite progress in other important crop plants, efforts to produce transgenic soybeans resistant to SMV have advanced slowly. In order to produce soybean lines that could confer pathogen derived resistance, plants were produced containing a coat protein gene and the 3′ UTR from SMV [25]. Coat protein gene transcripts were detected in transgenic lines and two of the soybean lines were highly resistant to infections with the SMV virus. These results represent the first example of stable genetically engineered SMV resistance in soybean.

The sense coat protein gene of soybean dwarf virus (SbDV) was used to acquire SbDV-resistant soybean plants [26]. These insertions were classified into two types: overexpression of SbDV-CP mRNA, or repression accumulation of SbDV-CP mRNA, and siRNA by RNA analysis prior to SbDV inoculation. In both cases, after infection with SbDV, most plants of these transgenic lines remained symptomless, contained little SbDV-specific RNA and exhibited SbDV-CP-specific siRNA. The possible mechanism of the achieved resistance was thought to be RNA silencing. This same group later used RNA silencing to create resistance for SbDV using inverted repeat-SbDV coat protein (CP) genes spaced by a β-glucuronidase sequence [27]. Upon infection with virus, transgenic plants showed no symptoms of the disease. Transgenic soybeans were shown to contain SbDV-CP-specific siRNA and little to no SbDV-specific RNA, suggesting that resistance to SbDV was achieved by an RNA silencing-mediated process.

3.3.3. Fungal

Fungi are the most common soybean pathogens and therefore represent targets for the development of disease resistant transgenic varieties in soybean. One of the more important fungal diseases affecting soybeans grown in the United States and Brazil is *Sclerotinia* stem rot (SSR) caused by the fungus *Sclerotinia sclerotiorum* (white mold). This mold has been associated with the presence of oxalic acid (OA). Treatment of plants with OA induced symptoms whereas metabolism of OA is correlated with fungal tolerance. Cunha et al., generated transgenic soybean lines that overexpressed oxalate decarboxylase (OXDC) [28]. When transgenic soybean lines were infected with white mold the disease progression showed significant reduction of severity that correlated with the level of transgene expression. Transgenic events expressing high levels of OXDC showed complete resistance demonstrating the feasibility of this approach.

Much of the research in the development of fungal disease resistance has focused on overexpression of a single gene to confer protection, though such a method favors co-evolution and pathogenic resistance. An alternative strategy taken by Li et al., was to create multigene resistance by overexpressing multiple anti-fungal genes [29]. Two such genes previously shown to be involved with fungal disease resistance are chitinase (CHI) and the barley ribosome-inactivating protein (RIP). While Li et al., successfully produced transgenic soybean overexpressing both traits, transgenic events were not challenged with fungal infection.

An alternative technology has shown promise with controlling fungal infection through the use of single-chain variable fragment (scFv) antibodies. While plants do not produce endogenous antibodies, they can express and correctly assemble antibody fragments. In fact, antibody production in soybean was first demonstrated in 1998 [30]. A similar antibody approach was recently taken by Brar and Bhattacharyya to control *Fusarium virguliforme* which is responsible for soybean sudden death syndrome (SDS) [31]. Using the pathogenic toxin Tox1 as a target, soybeans were transformed with an antibody gene encoding scFv anti-FvTox1 to create transgenic lines with enhanced foliar SDS resistance compared to control plants. Their results suggest that FvTox1 is a pathogenicity factor for the development of SDS and that expression of a soybean plant scFv antibody can reduce a toxin-induced plant disease. This biotechnology approach may be translational in fighting other plant diseases that are induced by pathogenic toxins.

To date there are no commercially available transgenic soybeans that confer resistance to disease, including fungal pathogens. In 2011 DuPont, Pioneer Hi-Bred, and Evogene announced a collaboration to develop soybean varieties displaying in-plant resistance to soybean rust [32]. This is a major step in the direction of creating the first commercially available transgenic soybean variety that is resistant to a fungal pathogen.

3.3.4. Nematode

Plant parasitic nematodes are a significant agricultural problem causing major limitations on crop yield and quality. It is estimated that plant parasitic nematodes cause approximately $157 billion [USD] in damage worldwide. Current approaches used to combat agricultural losses include the use of nematicides, cultivation techniques, and varieties with natural resistance. Nematicides include some of the most hazardous compounds used in agriculture and alternative control is required due to health and environmental concerns over their use. In soybean, the majority of yield loss can be attributed to infection by nematodes of the genus *Meloidogyne* and *Heterodera* commonly referred to as root knot nematodes (RKN) and soybean cyst nematodes (SCN), respectively. RKN and SCN infect plant roots and induce the formation of specialized feeding sites. The establishment and maintenance of feeding sites are crucial to the survival of nematodes making them an obvious target of interest for novel control strategies. One approach that has emerged in recent years is the use of *in planta* RNA interference (RNAi) to target genes of feeding nematodes. Through biotechnology, plants can be engineered to produce dsRNAs that silence essential nematode genes. Ingestion of plant-derived dsRNAs by the feeding nematode would trigger the RNAi process thereby inactivating targeted genes and preventing or limiting nematode infection. There are numerous genes known to be essential for nematode survival, and they have been the subject of past reviews [33, 34]. Many of these appear to be candidates for use in an *in planta* RNAi strategy to control nematode infection.

Steeves et al., was one of the first to demonstrate efficacy of an RNAi-based strategy to control SCN [35]. Transgenic soybeans were generated following transformation with an RNAi expression vector containing inverted repeats of a cDNA clone of the SCN major sperm protein (MSP). RNA silencing was elicited in the cyst nematode following ingestion of dsRNA molecules, and resulted in ~75% suppression of reproductive capabilities. Several years later Li et al., used RNAi to test potential gene targets known to be involved with nematode reproduction and fitness [36]. Soybean roots expressing small interfering RNAs against the SCN genes Cpn-1, Y25, and Prp-17 showed a significant reduction in transcript levels in nematode feeding sites. Furthermore, nematode suppression levels were similar to those observed with conventional resistance. Recently RNAi was used to disrupt genes involved with RKN gall formation [37]. Genes encoding tyrosine phosphatase (TP) and mitochondrial stress-70 protein precursor (MSP) were stably expressed in soybean roots, and following infection with RKN the number of galls was decreased by >90%. Nematode growth within roots was measured and the diameter of nematodes inside transformed soybean roots was reduced 5-fold over that of nematodes inside control roots.

Although there are a few cultivars of soybean that have natural resistance to some species of nematode, there are currently no commercially available soybean varieties that offer genetically modified resistance to nematodes. Over the past 10 years, there have been numerous candidate genes found within the nematode-plant interaction that hold potential for the development of novel genetically modified soybeans using an RNAi-based strategy. Results from the above studies show the potential of RNAi technology for reducing gall formation, limiting nematode reproduction and infection, and ultimately broadening soybean resistance to SCN and RKN. The production and eventual commercialization of nematode resistant soybean will benefit both producers and consumers by decreasing dependence on hazardous nematacides and increasing overall soy grain yield.

4. Soybean trait enhancements

In 2008 Monsanto announced their Sustainable Yield Initiative - a pledge to double the yields of corn, cotton, and soybeans by the year 2030 while simultaneously reducing aggregate key inputs such as water, land, and energy. While this will be an especially difficult task given that the vast majority of high-quality farm land is already in use, several recent reports involving transgenic soybean technologies support the notion that future biotechnological advances will indeed be able to help achieve such goals.

4.1. Crop yield

Crop yield is a highly complex trait, and increases in yield have previously been accomplished through a variety of methods involving traditional breeding and modern biotechnology. The introduction of transgenic crops in 1996 helped improve grain yield by protecting plants from insects and disease pathogens that often result in yield pressure if not treated. While new varieties of soybean combine the latest advances in both modern breeding with genetic modification technologies, there continues to be a search for gene-based approaches with potential to increase soy grain yield. Preuss et al., recently performed a large-scale screening for such yield increasing genes and reported that constitutive expression of an *Arabidopsis thaliana* B-box domain gene (BBX32) resulted in plants with increased plant height, node, flower, pod, and total seed number [38]. More importantly, field grown events showed a 5-8% increase in plant height, 8-10% increase in pod number, and 11-14% increase in total yield relative to control plants. It is believed that overexpression of AtBBX32 modulated circadian clock gene transcripts leading to an increase in the duration of reproductive developmental stages (R3 through R7) of the seed which presumably accounted for the increase in seed yield. Over the next decades, it is likely that seed varieties containing these and other yield traits will be commercialized.

4.2. Drought resistance

Drought is a major abiotic stress factor since it can greatly impact crop productivity and grain yield. Soybeans have developed several adaptive traits to endure periods of dry weather and drought. Inclusion of these traits into quality germplasm continues to be a major goal of traditional and marker-assisted breeding programs. While the genetic basis of drought tolerance is not well understood, researchers have focused on understanding physiological responses associated with drought (i. e. leaf wilting, water use efficiency, nitrogen fixation, and root growth biomass). While overexpression of single downstream gene targets have shown potential for increasing drought tolerance in *Arabidopsis* and tobacco model systems, the majority of these findings have not yet been translated to major crop species. One exception involves the overexpression of an endoplasmic reticulum-resident molecular chaperone binding protein (BiP) which is believed to regulate Ca2+ signaling responses. Valente et al., showed that BiP-overexpressing soybean lines exhibited decreases in leaf wilting, leaf water potential, and stomatal closure under reduced and deprived water conditions [39]. Furthermore, transgenic plants showed decreased rates of photosynthesis and transpiration, steady levels of osmolytes and dry root weight, decreased induction of drought-associated mRNAs, and delayed leaf senescence relative to control plants. While overexpression of BiP shows great potential as a target for increasing drought resistance, it will be important to compare grain yields in field-grown transgenic and control lines.

4.3. Increased oil content

Over the past decade, there has been a growing trend for industrial applications utilizing soybean oil, and these applications compete with those used for edible consumption. One example is the recent spike in soy-based biodiesel production which consumed just over 1 billion gallons of soybean oil in 2011 compared with 5 million gallons in 2001 [40]. The growing demand for soybean oil has sparked an interest in novel technologies that could be used to increase the relative oil content of soybean seeds. The retooling of soybean metabolism to increase oil content is not a simple task given that the absolute levels of seed oil and seed protein seem to be set. Increasing oil content comes at the expense of decreasing protein content, and vice versa. To date, only a few papers have reported successes in this area, and both involved manipulation of enzymes and substrate pools in the Kennedy pathway which is responsible for the production of triacylglycerols (TAGs) - the major component of soybean seed oil. In 2008, Lardizabel et al., overexpressed fungal diacylglycerolactetyltransferase (DGAT2) in soybean seeds [41]. DGAT2 converts diacylglycerols (DAGs) to TAGs. Transgenic soybeans overexpressing DGAT2 were grown at 63 locations within the United States and Argentina over five growing seasons, and showed a 1.5% increase in total seed oil with no reduction of seed protein content or yield. In 2009, Rao and Hildebrand overexpressed the yeast sphingolipid compensation (SLC1) protein in soybean seeds [42]. SLC1 has been shown to have lysophosphatidic acid acyltransferase (LPAT) activity which plays a role in the conversion of lysophosphatidic acid to phosphatidic acid, the precursor to DAG in the Kennedy pathway. Overexpression of yeast SLC1 resulted in soybean somatic embryos with 3. 2% increased oil content and stable transgenic lines with 1.5% increased oil content in seeds. Given current commodity pricing for soybean oil [43], a 1.5% increase in oil adds ~$1.2 billion [USD] in value to the United States soybean crop alone. As soybean oil prices rise it is anticipated that other metabolic engineering strategies will be developed and used to obtain similar increases in seed oil content.

5. Health and nutrition

Soybean is considered a dual use crop since it is a valuable source of seed protein and seed oil. Due to consumer health awareness and an increased demand for vegetable oil, much attention has been drawn to seed oil quality and content. To address these needs, efforts have been made with soybean to increase oxidative stability of soybean oil, enhance fatty acid content of oil, and increase total oil content within the seed. Significant progress has been made by breeders to improve overall yield of soybean, however minimal advancements have been made in the development of high-yield germ lines that have a major shift in carbon flux for increased total protein or oil content in the seed. This may be attributed to the inverse correlation between absolute oil and protein contents within soybean seeds. Biotechnology offers new tools for the development of soybeans that have improved oil quality for use in food, feed, and industrial applications. These nutritional enhancements have been achieved by directed modification of fatty acid biosynthesis to alter amounts of fatty acids that naturally occur in soybean, or to produce novel fatty acids. Two fatty acid profiles that have been targeted through genetic strategies include soybeans with low linolenic acid oil content and high oleic acid oil content.

5.1. Fatty acid content

Linolenic acid (LA) accounts for 10-13% of the total fatty acid content of soybean oil. This fatty acid reduces oxidative stability of oil which results in rancidity and decreased shelf-life. A family of three desaturase genes (*Gm*FAD3) contribute to LA biosynthesis in soybean. Flores et al., employed a targeted gene silencing approach to suppress the *Gm*FAD3 gene family using a single RNAi construct [44]. The down regulation of this gene family resulted in low linolenic soybeans with LA contents below 2%. Oleic acid (OA) is a pre-cursor of LA and is considered a healthy source of fat. Conventional soybean oil contains ~18% OA. While high oleic soybean oil has obvious nutritional value, conventional breeding of high oleic soybean lines have not materialized, in part due to the decrease of yield and environmental instability associated with bred traits. The observed yield drag may be attributed to an alteration in the fatty acid profiles within vegetative tissues of soybean. Progress toward decreasing this yield drag has been made by down regulating Δ^{12} desaturases (FAD2-1A and -1B) which converts OA to LA [45, 46]. Seeds resulting from this genetic approach have an increased OA content to ~80% without alterations of vegetative tissue fatty acid contents. In combination with high levels of OA, elevated steric acid levels in soybean oil is also desirable to meet needs of confectionary applications, and development of such soybeans may be possible in the future using similar approaches.

Nutritional enhancements such as increased ω-3 fatty acid levels are also desired for human food and animal feed consumption. Two fatty acids, γ-linolenic acid (GLA) and stearidonic acid (SDA) are of particular interest since they exhibit pharmacological properties and nutritional value, respectively. Sato et al., were successful at increasing these important fatty acids in soybeans by overexpressing a borage Δ^6desaturase which converts LA and α-linolenic (ALA) to GLA and SDA, respectively [47]. Field studies with transgenic soybean harboring the borage Δ^6desaturase produced GLA to ~27% and SDA to ~3% in seed oil [48]. To increase SDA levels in soy, Eckert et al., pyramided the borage Δ^6desaturase with an *Arabidopsis* Δ^{15}desaturase which converts LA to ALA [49]. It was reasoned that increased pools of ALA would lead to increased levels of SDA via the Δ^6desaturase. This strategy resulted in soybean lines with 21.6% SDA when grown under greenhouse conditions. SDA in turn is a precursor for the long chain polyunsaturated fatty acids eicosapentaenoic acid (EPA) and docosahexaenoic acid (DHA). Diets rich in these fatty acids are associated with cardiovascular fitness. While humans and animals possess the enzymatic machinery to convert SDA to EPA and DHA, diets are often supplemented with EPA and DHA derived from fish oil since vascular plants lack the genes for synthesis of these fatty acids. Attempts have been made to assemble these pathways in soybean by using biosynthetic genes from fungi, algae, and protists along with seed specific co-expression of these genes [50]. In such experiments EPA levels approached 20% and DHA levels represented up to 3% of total seed fatty acid content in soybean. The production of such important fatty acids has significant market potential, particularly with respect to human, poultry, pet, and aquaculture feed applications.

5.2. Tocopherols

Tocopherols are lipid-soluble antioxidants that are extracted during the commercial processing of soybean seeds and add to the stability of the oil. In soybean there are four forms of tocopherols (α, β, γ and δ,) classified by the number of methyl groups present on the molecule. These molecules are collectively referred to as vitamin E, with δ - and γ -tocopherols being the most predominant forms in seeds. Biotechnological enhancements of tocopherols in soybean have mainly focused on increasing the amounts of α-tocopherol since this form has the highest nutritional value. Attempts to increase total levels of tocopherols have only been marginally successful. Examples of studies to increase total levels have focused on up-regulating homogentisatephytyltransferase (HPT) activity in seeds. While HPT catalyzes the first step in tocopherol synthesis, overexpressing lines resulted in little increase of tocopherol in transgenic seeds [51, 52]. A modest increase in tocopherol content was achieved by the expression of corresponding HPT genes from *Arabidopsis* and *Synechocystis* with a strong seed specific promoter [52]. This approach resulted in a 1.5-fold increase of total tocopherol content. This same group also expressed a bacterial chorismatemutase-prephenate dehydrogenase (TYRA) gene with several other enzymes under the control of seed specific promoters and observed a >10- fold increase in total vitamin E type molecules.

To generate soybean lines with increased levels of tocopherols a more direct approach has involved overexpression of homogentisategeranylgeranlytransferase (HGGT), an enzyme involved in the biosynthesis of tocopherols in monocots. Recently transgenic expression of rice HGGT was expressed in soybean with a seed-specific and constitutive promoter [53]. Transgenic soybean expressing the HGGT gene had significantly higher levels of antioxidant activities and showed enhanced vitamin E levels associated with the presence of all forms of tocopherols, including tocotrienols (with the exception of the β- form) which are not found naturally in soybean.

Enhanced amounts of α-tocopherol in soybean have proven easier to metabolically engineer. Expression of genes for two enzymes responsible for methylation of tocopherol head groups (VTE3 and VTE4) from *Arabidopsis* were co-expressed within the seed and generated plants with α-tocopherol levels greater than 90% of total tocopherol content [54]. The total levels of tocopherol remained the same in these seeds, showing a shift in tocopherol to mainly the α-form yielding a 5-fold increase in vitamin E activity. This research can lead to soybean oil with enhanced vitamin E and more nutritional value for consumers.

5.3. Dietary amino acids

Soy is also deficient in several essential dietary amino acids, most notably methionine and cysteine due to their high sulfur contents. Albumins from Brazil nuts, sunflowers, and corn have been expressed in soybean and although they resulted in increased methionine and cysteine levels, they are not adequate enough to avoid supplementation of these amino acids in animal feed and human diets. The physical synthesis of cysteine is carried out by the enzyme O-acetylserinesulfhydrylase (OASS). In an attempt to increase sulfur containing amino acids in soybean, Kim et al., overexpressed cytosolic OASS and found that transgenic seeds contained elevated levels of both protein bound cysteine (58-74%) and free cysteine (22-32%) [55]. Another approach used to increase sulfur amino acid content in soybean is the use of a maize zein gene which gives rise to several species of insoluble proteins containing high levels of methionine. Dinkins et al., overexpressed maize zein in soybean seeds and observed a 12-20% increase in methionine and 15-35% increase in cysteine without adverse effects on protein composition [56].

The essential amino acids lysine and threonine have also been explored for their potential to create nutritionally enhanced soybean. An increased level of lysine was observed by genetically engineering the lysine biosynthetic pathway to circumvent the normal feedback regulation of the enzymes aspartokinase and dihydrodipicolinic acid synthase in soybean [57]. In this case, a >100-fold increase of free lysine and 5-fold increase of total seed lysine content was observed. More recently, soybeans showing enhanced threonine levels have been engineered using seed-specific expression of lysine-insensitive variants of aspartate kinases from bacteria [58]. This strategy produced transgenic soybeans with a 100-fold increase in threonine levels and 3.5-fold increase in total free amino acid content without negative impacts on seed morphology or germination. While enhancement of essential amino acids in soybean seeds has clear potential for commercial applications, it will be important to demonstrate that transgenic soybean with increased nutritional enhancement traits also maintain optimal agronomic characteristics when grown in the field under a variety of conditions.

6. Soybeans as bioreactors for pharmaceuticals

Recombinant proteins are widely used in medicine, research laboratories, food and nutrition, and play a key role in important agriculture and biopharmaceutical industries. Since the development of recombinant DNA technology in the early 1970's, the commercial production of recombinant proteins has traditionally relied on a variety of protein expression systems, each with intrinsic advantages and disadvantages. Over the years several methods have been used to produce recombinant proteins. Traditionally, prokaryotic systems based on fermentation have been used for the production of biopharmaceuticals and enzymes. The bacteria *Escherichia coli*, being one of the earliest and most widely used host for this method, has been used in the production of human insulin since the 1970's [59]. Other platforms include the use of fungal cells and yeast, which have been used as an expression system that is able to perform many of the post-translational modifications required by recombinant protein production. Recombinant proteins that require more complex modifications can be produced using insect or mammalian cells, or transgenic animals. However, major disadvantages associated with these platforms include the inability to perform complex post-translational modifications, the alteration in glycosylation patterns affecting protein activity, the high overall cost associated with manufacturing, the potential for contamination, and long time commitments associated with production in transgenic animals making these platforms impractical for the production of most proteins.

Over the past two decades, there has been a push for recombinant protein technologies to move towards more effective expression systems. These systems must be safe, cost-effective, and conducive to post-translational modifications and processing methods on a large scale. Transgenic plants represent an economical system for accurate expression of complex recombinant proteins on a large scale. Plant cells combine the potential for full post translational modifications and correct protein folding with simple growth requirements. The use of plants as a platform for recombinant protein production has a low risk of contamination with prions, viruses, and other pathogens that infect mammalian cells, and therefore offer advantages that are not associated with existing expression systems. An important advantage of plants as a bioreactor is that recombinant proteins and biopharmaceuticals may be expressed in multiple plant organs including seeds which naturally accumulate high amounts of stored proteins. In general, crops that have higher protein content are more cost-effective for molecular farming. Among recombinant systems that utilize seeds, soybeans present an exceptionally high endogenous protein content, which can reach up to 40% of the dry seed weight. Soybeans are an ideal source of protein for food and feed thus occupying a unique position as a premier target for genetic engineering, and as a platform for the production of recombinant protein. An important characteristic favoring expression in soybean seeds is that these organs have evolved as specialized compartments to store proteins for embryo nutrition. Based on this, soybean seeds offer an environment with metabolic adaptations that permit the stable and long-term storage of proteins, reducing the requirement for sophisticated and expensive conditions for storage. This makes it possible to stockpile harvested seeds so that the downstream processing can be made available based on the demands of the industry. Similarly, soybean seeds provide a compact compartmentalization biomass, which can considerably reduce overall production costs since purification expenses are typically inversely proportional to the final concentration in the plant biomass.

The concept of a soy-derived pharmaceutical was tested back in 1995 when Cho et al., developed a transformation expression cassette using a soybean seed-specific lectin promoter to test for potential expression of the β-glucuronidase reporter gene [60]. This same expression cassette was used to produce bovine β-casein in soybean which accumulated to 0.1-0.4% of seed total soluble protein (TSP) [61]. A follow-up paper characterized post-translational processing, subcellular localization to the PSV, and purification of transgenic β-casein [62]. These proof-of-concept studies showed that a seed-specific promoter could be used to target stable expression of proteins with commercial value in soybean seed.

6.1. Antibodies

Monoclonal antibodies (mAbs) have played a major role in the advancement of biotechnology and development of mAb-based therapeutics and diagnostics. Plants have great potential to serve as a platform for the production of antibodies for therapeutic use. One of the first reports of a functional plant-based antibody was developed in soybean. In an effort to explore cost effective methods of mucosal immunoprotection against sexually transmitted diseases, Zeitlin et al., expressed a humanized monoclonal anti-herpes simplex virus 2 (HSV-2) antibody in leaf tissue [63]. That study compared purified soy-derived and mammalian cell-derived HSV-2 mAbs and found that both were similar with respect to stability in human semen and cervical mucus over a 24 hour period. Both antibodies were also able to diffuse in human cervical mucus, and were efficacious in preventing vaginal HSV-2 infection in a murine model.

6.2. Vaccines

When plants are mentioned as a platform for the production of pharmaceuticals, the concept of edible vaccines often comes to mind. Edible vaccines are desirable since they would eliminate the use of needles and specialized personnel to administer shots, which may have broad applicability in developing nations. Soybean seeds represent an ideal target for the production of vaccines since soymilk-based formulations are safe and can be easily administered orally. Furthermore, soybean seeds are capable of storing vaccine antigens for many years at ambient temperatures without loss or degradation of the antigen [64-66]. Such features can reduce the need for a cold chain therefore reducing costs.

Vaccines that can be administered at mucosal surfaces offer systemic immunity. Subunit antigens used to vaccinate orally or nasally are often ineffective and require formulation with a mucosal adjuvant for increased efficacy. The heat labile toxin (LT) of *E. coli* is comprised of a single A subunit (LTA) and pentameric B subunit (LTB) and has been shown to act as both a strong mucosal adjuvant as well as an antigen [67]. Moravec et al., targeted LTB expression to the endoplasmic reticulum of seed storage parenchyma cells where it accumulated to levels up to 2.4% of seed TSP [68]. Mice orally immunized with seed extracts containing LTB induced both, systemic IgG and mucosal IgA anti-LTB antibody responses. The soybean derived LTB also increased an antibody response against a co-administered bacterial FimHt antigen by 500-fold demonstrating that soy-derived LTB may function as an oral adjuvant.

Several subunit antigens have been expressed in soybeans that are important to the agricultural industry and could lead to effective vaccines. FanC is a specialized adhesion protein located on the bacterial surfaces of Enterotoxigenic *E. coli* (ETEC). K99 and other ETEC strains cause acute diarrhea in humans and livestock and can be severe and even cause death if left untreated. ETEC vaccinations are routinely administered parenterally to pregnant farm animals in order to stimulate systemic immunity and offer protection in newborns. An edible form of this vaccine has the potential to increase efficacy by conferring mucosal immunity at sites of pathogen invasion. Piller et al., constitutively overexpressed the bacterial FanC antigen in soybeans and reported stable accumulation to levels representing ~0.4% TSP in both leaves and seeds [69]. Mice immunized with adjuvanted soymilk formulations containing FanC elicited FanC-specific systemic and cellular immune responses demonstrating immunogenicity of the soy-derived antigen.

In another study a soybean-based vaccine was developed against the virus that causes porcine reproductive and respiratory syndrome (PRRS) [70]. PRRS is a serious health problem among breeding swine herds and the current vaccine is not efficacious when applied in the field. Vimolmangkang et al., overexpressed a nucleocapsid protein (PRRSV-ORF7) that accumuolated to 0. 64% of seed TSP. Intragastric immunization of mice with transgenic seed extract, in the absence of adjuvant, induced specific humoral and mucosal immune responses against PRRSV-ORF7 [70].

6.3. Therapeutics

Protein therapeutic use is limited by the shortfalls in manufacturing capacity and the high cost of production. While an aging population is a key driver of the protein therapeutics market, the potential for future growth is dependent largely on the industry overcoming drug delivery challenges and cost issues. Plants are cost-effective systems that excel at producing complex therapeutic proteins and therefore could help address some of these issues. The high protein content of soybean seeds, low costs associated with growth, simplified purification methods, and safety, make soybean a unique platform for the production of protein-based therapeutics.

Russell et al., expressed human growth hormone (hGH) in soybean with transformation cassettes using both the constitutive 35S promoter as well as a soybean seed-specific promoter 7S β-conglycyinin [71]. The resulting expression of hGH, both constitutively and within the seed, was detected at low levels of 0.0008% TSP. More recently hGH was expressed in soybean seeds with a more effective expression cassette utilizing the 7S α' subunit of β-conglycinin promoter and α-coixin signal peptide. In this case, hGH was directed to protein storage vacuoles within the seed and accumulated to 2.9% TSP. Bioassays demonstrated that the soy-derived hGH was fully active [66]. The cost of recombinant *E. coli*-derived hGH is still a very expensive therapy. Having such a high level of bioactive hGH protein expression in soybean seeds demonstrates the potential for high-yield production of recombinant proteins in soybean seeds and could lead to reduce costs for large-scale production of therapeutic molecules.

Human basic fibroblast growth factor (bFGF) is another high value therapeutic that has been expressed in soybean seeds [72]. This therapeutic was expressed under the control of the soybean seed specific G1 promoter and endogenous signal sequence from soybean. The bFGF protein accumulated to levels of ~2.3% of seed TSP and biological activity of the transgenic protein was confirmed by its mitogenic activity in mice.

Recombinant expression of Insulin was first reported using E. coli [59] and has since been commercialized. Like many pharmaceuticals derived from other expression systems, the potential for contamination along with high costs associated with production remain considerable for this hormone. To show that a soybean expression system could address some of these issues Cunha et al., used a sorghum γ-kafirin seed storage protein promoter and α-coixin PSV signal peptide to target recombinant proinsulin expression to soybean seeds [65]. Transgenic protein was stably expressed in seeds though accumulation levels were not reported. Transgenic seeds containing proinsulin were stable for up to seven years when stored under ambient storage conditions.

The soybean platform has also been used to produce a therapeutic for reducing systolic blood pressure. Novokinin is a hypotensive peptide that has vasorelaxing activity [73]. Novakinin was expressed in soybean seeds under the control of a modified β-conglycinin promoter and accumulated to 0.5% of seed TSP. A purified soy-derived formulation, as well as a less pure defatted flour formulation, was orally administered to groups of spontaneously hypertensive rats. Both the purified and partially purified formulations successfully reduced systolic blood pressure after a single dose [74].

Haemophilia B is a bleeding disorder that results from a deficiency of human coagulation factor IX (hFIX). The current treatment for this disease is intravenous infusion of plasma-derived or recombinant hFIX protein. While this treatment is effective at preventing and arresting hemorrhage, it is very costly and the protein is difficult to produce in large quantities. Using a biolistic transformation approach, hFIX expression was targeted to soybean seeds using the soy 7S promoter and coixin signal peptide [75]. Recombinant hFIX protein accumulated to 0.23% of seed TSP, and purified protein exhibited blood-clotting activity up to 1. 4% of normal plasma demonstrating functionality and efficacy of the soy-derived protein. The recombinant protein was stable for 6 years when stored at room temperature.

Soybeans are also capable of supporting expression and stable accumulation of large and complex proteins that can be difficult or impossible to express using current expression systems. Human thyroglobulin (hTG) is a 660 kDa homodimeric protein that is used as a protein standard and diagnostic for the detection of thyroid disease. To date, no expression system has been capable of producing a recombinant form of hTG which is likely due to strict requirements for correct post-translational modification and proper folding during protein synthesis. As a result, commercial hTG supplied to manufacturers for their assay kits is derived from cadaver and surgically removed thyroid tissue. The heterogeneity and lack of uniformity of commercially-purified hTG preparations is a major factor of variation between kits of different manufacturers. To explore the potential of soybean as a platform for production of large and complex proteins, Powell et al., used the 7S promoter and endogenous hTG signal peptide to target recombinant expression of hTG to soybean seeds [76]. Transgenic lines showed stable expression of full length hTG dimeric protein over multiple generations, and accumulated the protein to levels approaching 1.5% of seed TSP. Functionality of soy-derived hTG was demonstrated with commercial ELISA kits developed specifically for the detection of hTG in patient sera. The expression of 660 kDa dimerich TG appears to be the largest functional recombinant protein expressed in any plant system to date, and demonstrates the practicality of soy as an alternative system for the expression of proteins that are recalcitrant to expression in traditional systems [76].

Soybeans have a high intrinsic capacity for protein production and storage. Other than variability caused by nutrient modulation or environmental effects, the relative distribution of seed protein is primarily determined by genetics. Several groups have been able to achieve recombinant protein expression in soybean seeds at respectable levels (approaching 3% of seed TSP). However, knowledge of protein distribution in soybeans may help to further maximize expression levels. Schmidt and Herman tested this theory by overexpressing Green Fluorescent Protein (GFP) in soybeans with a β-conglycinin suppression background, to observe whether proteome rebalancing would result in a higher GFP yield [77]. They found that the rebalancing of intrinsic proteins could be exploited to obtain protein yields which increased ~4-fold in suppression backgrounds and approached levels representing >7% seed TSP. Thus, proteome rebalancing may represent a strategy that can be used to develop soybean lines capable of producing high levels of recombinant proteins in the future.

7. Conclusion

Over the past 25 years, soybean production in the United States has grown by nearly 57% while during that same period the number of acres used to grow soybeans has increased by only 24%. It is predicted that the worldwide requirement for grain will rise by 40-70% by the year 2050, driven in large part by the growing world population and the increase in demand for protein-rich diets. Clearly the demand for soybean protein and soybean oil is outpacing grain production, which in turn is outpacing available land for growing soybeans. Over the next 50 years, farmers will need to produce as much food as they did in the previous 10,000 years combined, and with fewer resources. The identification of various agricultural improvements such as herbicide, insect, and disease resistance which will allow farmers to obtain increased yields with reduced environmental inputs will be crucial. Traits that not only increase grain yield, but also improve the absolute levels of soy protein and soy oil within the seed will also be important for producers and consumers worldwide. While soybean is recognized for its high protein content, it is also the most widely grown oil-seed crop in the United States. Enhancing nutritional value of soybean oil will greatly increase the effectiveness and value of soy as a food crop, help meet the needs of a growing population, and improve human health.

Over the past decade, soybean has emerged as an ideal expression platform with potential to address current healthcare needs. These unmet needs include cost-effective alternatives to existing protein-based therapeutics, simplified methods for the administration of therapeutics, and the development of reagents that could lead to better diagnostic assays and novel medical devices. Soybeans are unique with respect to protein expression platforms. They are safe to consume, cost-effective to grow, rich in protein content, and stable for years under ambient storage conditions. They have been engineered to express a variety of potential therapeutics, including mAbs, vaccine antigens and adjuvants, hormones, growth factors, and blood-clotting factors. Seed-based expression of 660 kDa homodimerich TG underscores the potential of the soybean system to produce large and complex proteins that cannot be produced in yeast, insect, and mammalian cell cultures. The efficacy of engineered therapeutics in crude soymilk formulations could lead to oral vaccines and therapies that require little, if any, purification from other seed proteins. Reports demonstrating long-term stability of seed-derived therapeutics in the absence of climate control directly address cold chain issues associated with vaccines and other therapeutics. With seed protein levels of ~40% and transgene expression levels approaching 3% of TSP, a single soybean plant yielding 300 seeds can produce >500 mgs of transgenic protein. To put this in perspective, a single soybean plant can produce 500 doses of a vaccine antigen administered at 1 mg/ml, or alternatively, $50,000 [USD] of a therapeutic valued at $100/mg [USD].

The use of soybean as a platform for the production of therapeutics represents a technology with the potential to revolutionize our current approaches to healthcare. Harnessing the full potential of the soybean platform will depend on further increasing stable transgene expression levels, developing efficient purification methods, obtaining interest from pharmaceutical partners, and overcoming issues associated with commercialization. The production of vaccines, antibodies, and other therapeutic proteins will undoubtedly continue to develop over the next decade. As biotechnology evolves, so does the role of soybean - from the field to the bedside.

Glossary

EPSPS:5-enolpyruvylshikimate-3-phosphate synthase

PAT:Phosphinothricin-N-acetyltransferase

ALS:Acetolactate synthase

DMO: Dicambamonooxygenase

HPPD:Hydroxyphenylpyruvatedioxygenase

GM:Genetically modified

Bt:*Bacillus thuringiensis*

Cry1:Crystalline proteins

QTL:Quantitative trait loci

BPMV:Bean pod mottle virus

SMV:Soybean mosaic virus

SbDV:Soybean dwarf virus

SSR:*Sclerotinia* stem rot

OA:Oxalic acid

OXDC:Oxalate decarboxylase

CHI:Chitinase

RIP:Ribosome-inactivating protein

scFv:Single-chain variable fragment

SDS:Sudden death syndrome

RKN:Root knot nematodes

SCN:Soybean cyst nematodes

RNAi:RNA interference

MSP:Major sperm protein

TP:Tyrosine phosphatase

TAGs:Triacylglycerols

DAGS:Diacylglycerols

SCL1:Yeast sphingolipid compensation protein

LPAT:Lysophosphatidic acid acyltransferase

LA:Linolenic acid

OA:Oleic acid

GLA:γ-linolenic acid

SDA:Stearidonic acid

ALA:α-linolenic

EPA:Eicosapentaenoic acid

DHA:Docosahexaenoic acid

HPT:Homogentisatephytyltransferase

HGGT:Homogentisategeranylgeranlytransferase

OASS:O-acetylserinesulfhydrylase

TSP:Total soluble protein

PSV:Protein storage vacuole

mAbs:Monoclonal antibodies (mAbs)

HSV-2:Herpes simplex virus 2

LT:Heat labile toxin of *E. coli*

ETEC:Enterotoxigenic*E. coli*

PRRS:Porcine reproductive and respiratory syndrome

hGH:Human growth hormone

bFGF:Human basic fibroblast growth factor

hFIX:Human coagulation factor IX

hTG:Human thyroglobulin

GFP:Green Fluorescent Protein

Author details

Laura C. Hudson, Kevin C. Lambirth, Kenneth L. Bost and Kenneth J. Piller

University of North Carolina at Charlotte and SoyMeds, Inc., USA

References

[1] Horsch RB, Fry JE, Hoffmann NL, Eichholtz D, Rogers SG, Fraley RT. A simple and general method for transferring genes into plants. Science. 1985;227:1229–1231.

[2] Hinchee MAW, Conner-Ward DV, Newell CA, McDonnell RE, Sato SJ, Gasser CS, Fischhoff DA, Re DB, Fraley RT, Horsch RB. Production of transgenic soybean plants using *Agrobacterium*-mediated DNA transfer. Nat. Biotechnol. 1988;6:915–922.

[3] Parrott WA, Williams EG, Hildebrand DF, Collins GB. Effect of genotype on somatic embryogenesis from immature cotyledons of soybean. Plant Cell Tissue Organ Cult. 1989;16:15–21.

[4] Trick HN, Finer JJ. Sonication-assisted *Agrobacterium* mediated transformation of soybean [*Glycine max* (L.) Merrill] embryogenic suspension culture tissue. Plant Cell Rep. 1998;17:482–488.

[5] Christou P, McCabe DE, Swain WF. Stable transformation of soybean callus by DNA-coated gold particles. Plant Physiol. 1988;87:671–674.

[6] McCabe DE, Swain WF, Martinell BJ, Christou P. Stable transformation of soybean (*Glycine max*) by particle acceleration. Nat. Biotechnol. 1988;6:923–926.

[7] Finer JJ, McMullen MD. Transformation of soybean via particle bombardment of embryogenic suspension culture tissue. In Vitro Cell. Dev. Biol. 1991;27P:175–182.

[8] Aragão FJL, Sarokin L, Vianna GR, Rech EL. Selection of transgenic meristematic cells utilizing a herbicidal molecule results in the recovery of fertile transgenic soybean [*Glycine max* (L.)Merril] plants at a high frequency. Theor. Appl. Genet. 2000;101:1–6.

[9] Franz JE, Mao MK, Sikorski JA. Glyphosate: A Unique Global Pesticide. Washington, DC: American Chemical Society. 1996.

[10] Bayer Crop Science. http://www.bayercropscience.com.

[11] Mathesius CA, Barnett JF Jr, Cressman RF, Ding J, Carpenter C, Ladics GS, Schmidt J, Layton RJ, Zhang JX, Appenzeller LM, Carlson G, Ballou S, Delaney B. Safety assessment of a modified acetolactate synthase protein (GM-HRA) used as a selectable marker in genetically modified soybeans. RegulToxicolPharmacol. 2009;55(3): 309-320.

[12] Lee YT, Duggleby RG. Mutagenesis studies on the sensitivity of *Escherichia coli*acetohydroxyacid synthase II to herbicides and valine. Biochem. J. 2000;350:69-73.

[13] Behrens MR, Mutlu N, Chakraborty S, Dumitru R, Jiang WZ, LaVallee BJ, Herman PL, Clemente TE, Weeks DP. Dicamba Resistance: Enlarging and preserving biotechnology-based weed management strategies. Science 25 2007;316(5828):1185-1188.

[14] vanAlmsick A. HPPD-Inhibitors - A proven mode of action as a new hope to solve current weed problems. Outlooks on Pest Management. 2009;20(1);27-30.

[15] Parrott WA, All JN, Adang MJ, Bailey MA, Boerma HR, Stewart CN. Recovery and evaluationof soybean plants transgenic for a *Bacillus thuringiensis* var. *kurstaki* insecticide gene. In Vitro Cell. Dev.Biol. 1994;30P:144-149.

[16] Stewart CN, Adang MJ, All JN, Boerma HR, Cardineau G, Tucker D, Parrott WA. Genetic transformation, recovery, and characterization of fertile soybean transgenic for a synthetic *Bacillus thuringiensis* cryIAc gene. Plant Physiol. 1996; 112(1): 121–129. \

[17] Walker DR, All JN, Mcpherson RM, Boerma HR, Parrott WA. Field evaluation of soybean engineered with a synthetic cry1Ac transgene for resistance to corn earworm, soybean looper, velvetbean caterpillar (Lepidoptera: Noctuidae),and lesser cornstalk borer (Lepidoptera: Pyralidae). J Econ Entomol. 2000;93(3);613- 622.

[18] Mcpherson RM, MacRae TC. Evaluation of transgenic soybean exhibiting high expression of a synthetic *Bacillus thuringiensis* cry1A transgene for suppressing lepidopteran population densities and crop injury. J Econ Entomol. 2009;102(4)1640-1648.

[19] Cregan PB, Jarvik T, Bush AL, Shoemaker RC, Lark KG, Kahler AL, Kaya N, vanToai TT, Lohnes DG, Chung J, Specht JE. An integrated genetic linkage map of the soybean genome. Crop Science. 1999;39:1464-1490.

[20] Rector BG, All, JN, Parrott WA, Boerma HR. Quantitative trait loci for antibiosis resistance to corn earworm in soybean. Crop Science. 2000;40:233-238.

[21] Walker DR, Narvel JM, Boerma HR, All JN, Parrott WA. A QTL that enhances and broadens Bt insect resistance in soybean. TheorAppl Genet. 2004;109(5):1051-1057.

[22] Monsanto. http:// www.monsanto.com.

[23] Di R, Purcell V, Collins GB, Ghabiral SA. Production of transgenic soybean lines expressing the bean pod mottle virus coat protein precursor gene. Plant Cell Reports. 1996;15:746-750.

[24] Reddy MS, Ghabrial SA, Redmond CT, Dinkins RD and Collins GB. Resistance to Bean mod mottle virus in transgenic soybean lines expressing the capsid polyprotien. Phytopathology. 2001;91:831-838.

[25] Wang X, Eggenberger AL, Nutter FW Jr., Hill JH. Pathogen-derived transgenic resistance to soybean mosaic virus in soybean. Molecular Breeding. 2001;8(2):119-127.

[26] Tougou M, Furutani N, Yamagishi N, Shizukawa Y, Takahata Y, Hidaka S. Development of resistant transgenic soybeans with inverted repeat-coat protein genes of soybean dwarf virus. Plant Cell Rep. 2006;25(11):1213-1218.

[27] Tougou M, Yamagishi N, Furutani N, Shizukawa Y, Takahata Y, Hidaka S. Soybean dwarf virus-resistant transgenic soybeans with the sense coat protein gene. Plant Cell Rep. 2007;26(11):1967-1975.

[28] Cunha WG, Tinoco MLP, Pancoti HL, Ribeiro RE, Aragão FJL. High resistance to *Sclerotinia sclerotiorum* in transgenic soybean plants transformed to express an oxalate decarboxylase gene. Plant Pathology. 2010;59(4):654-660.

[29] Li HY, Zhu YM, Chen Q, Conner RL, Ding XD, Li J, Zhang BB. Production of transgenic soybean plants with two anti-fungal protein genes via *Agrobacterium* and particle Bombardment. Biologia Plantarum. 2004;48(3):367-374.

[30] Zeitlin L, Olmsted SS, Moench TR, Co MS, Martinell BJ, Paradkar VM, Russell DR, Queen C, Cone RA, Whaley KJ. A humanized monoclonal antibody produced in transgenic plants for immunoprotection of the vagina against genital herpes. Nature Biotechnol. 1998;16:1361–1364.

[31] Brar HK, Bhattacharyya MK. Expression of a single-chain variable-fragment antibody against a *Fusarium virguliforme* toxin peptide enhances tolerance to sudden death syndrome in transgenic soybean plants. MPMI. 2012;25(6);817–824.

[32] Evogen. http://www.evogene.com/News-Events/

[33] Gheysen G, Vanholme B. RNAi from plants to nematodes. Trends Biotechnol. 2007;25(3):89-92.

[34] Lilley CJ, Bakhetia M, Charlton WL, URWIN PE. Recent progress in the development of RNA interference for plant parasitic nematodes. Molecular Plant Pathology. 2007;8:701–711.

[35] Steeves RM, Todd TC, Essig JS, Trick HN. Transgenic soybeans expressing siRNAs specific to a major sperm protein gene suppress *Heterodera glycines* reproduction. Funct Plant Biol. 2006;33:991–999.

[36] Li J, Todd TC, Oakley TR, Lee J, Trick HN. Host-derived suppression of nematode reproductive and fitness genes decreases fecundity of *Heterodera glycines Ichinohe*. Planta. 2010;232(3):775-85.

[37] Ibrahim HM, Alkharouf NW, Meyer SL, Aly MA, Gamal El-Din Ael K, Hussein EH, Matthews BF. Post-transcriptional gene silencing of root-knot nematode in transformed soybean roots. Exp Parasitol. 2011;127(1):90-99.

[38] Preuss SB, Meister R, Xu Q, Urwin CP, Tripodi FA, et al. Expression of the *Arabidopsis thaliana* BBX32 gene in soybean increases grain yield. PLoS ONE. 2012;7(2):e30717.

[39] Valente MA, Faria JA, Soares-Ramos JR, Reis PA, Pinheiro GL, Piovesan ND, Morais AT, Menezes CC, Cano MA, Fietto L.G. et al. The ER luminal binding protein (BiP) mediates an increase in drought tolerance in soybean and delays drought-induced leaf senescence in soybean and tobacco J. Exp. Bot. 2009;60(2):533-546.

[40] Soystats. http://www.soystats.com.

[41] Lardizabal K, Effertz R, Levering C, Mai J, Pedroso MC, Jury T, Aasen E, Gruys K, Bennett K.Expression of *Umbelopsis ramanniana*DGAT2A in seed increases oil in soybean. Plant Physiol. 2008;148:89–96.

[42] Rao, SS, Hildebrand D. Changes in oil content of transgenic soybeans expressing the yeast SLC1 gene. Lipids. 2009;44(10):945-951.

[43] CME Group. http://www.cbot.com.

[44] Flores T, Karpova O, Su X, Zeng P, Bilyeu K, Sleper DA, Nguyen HT, Zhang ZJ. Silencing of GmFAD3 gene by siRNA leads to low alpha-linolenic acids (18:3) of fad3-mutant phenotype in soybean [*Glycine max* (Merr.)]. Transgenic Res. 2008;17(5): 839-850.

[45] Mazur B, Krebbers E, Tingey S. Gene discovery and product development for grain quality traits. Science. 1999;285: 372–375.

[46] Buhr T, Sato S, Ebrahim F, Xing A, Zhou Y, Mathiesen M, Schweiger B, Kinney AJ, Staswick P, Clemente T. Ribozyme termination of RNA transcripts down-regulate seed fatty acid genes in transgenic soybean. Plant J 2002;30:155–163.

[47] Sato S, Xing A, Ye X, Schweiger B, Kinney A, Graef G, Clemente T. Production of g-linolenic acid and stearidonic acid in seeds of marker free transgenic soybean. Crop Sci. 2004;44:646–652.

[48] Clement T, and Calhoon EB. Soybean Oil: Genetic Approaches for Modification of Functionality and Total Content. Plant Physiology. 2009;151(3):1030-1040.

[49] Eckert H, LaVallee BJ, Schweiger BJ, Kinney AJ, Cahoon EB, Clemente T. Co-expression of the borage Δ^6 desaturase and the *Arabidopsis* Δ^{15}desaturase results in high ac-

cumulation of stearidonic acid in the seeds of transgenic soybean. Planta. 2006;224:1050–1057.

[50] Kinney AJ, Cahoon EB, Damude HG, Hitz WD, Kolar CW, Liu ZB. Production of very long chain polyunsaturated fatty acids in oilseed plants. World Patent Application. 2004.

[51] Savidge B, Weiss JD, Wong YH, Lassner MW, Mitsky TA, Shewmaker CK, Post-Beittenmiller D, Valentin HE. Isolation and characterization of homogentisatephytyl-transferase genes from *Synechocystis* sp. PCC 6803 and Arabidopsis. Plant Physiol. 2002;129: 321–332.

[52] Karunanandaa B, Qi Q, Hao M, Baszis SR, Jensen PK, Wong YH, Jiang J, Venkatramesh M, Gruys KJ, Moshiri F, Post-Beittenmiller D, Weiss JD, Valentin HE. Metabolically engineered oilseed crops with enhanced seed tocopherol. Metab Eng. 2005;7(5-6):384-400.

[53] Kim YH, Lee YY, Kim YH, Choi MS, Jeong KH, Lee SK, Seo MJ, Yun HT, Lee CK, Kim WH, Lee SC, Park SK, Park HM. Antioxidant activity and inhibition of lipid peroxidation in germinating seeds of transgenic soybean expressing OsHGGT. J Agric Food Chem. 2011;59(2):584-591.

[54] Van Eenennaam AL, Lincoln K, Durrett TP, Valentin HE, Shewmaker CK, Thorne GM, Jiang J, Baszis SR, Levering CK, Aasen ED, Hao M, Stein JC, Norris SR, Last RL. Engineering vitamin E content: from *Arabidopsis* mutant to soy oil. Plant Cell. 2003;15(12):3007-3019.

[55] Kim WS, Chronis D, Juergens M, Schroeder AC, Hyun SW, Jez JM, Krishnan HB. Transgenic soybean plants overexpressing O-acetylserinesulfhydrylase accumulate enhanced levels of cysteine and Bowman-Birk protease inhibitor in seeds. Planta. 2012;235(1):13-23.

[56] Dinkins RD, Reddy MSS, Meurer CA, Yan B, Trick H, Thibaud-Nissen F, Finer JJ, Parrott WA, Collins GB. Increased sulfur amino acids in soybean plants overexpressing the maize 15 kDazein protein. In Vitro Cell. Devel. Biol. Plant. 2001;37(6):742-747.

[57] Falco SC, Guida T, Locke M, Mauvais J, Sanders C, Ward RT, Webber P. Transgenic canola and soybean seeds with increased lysine. Biotechnology. 1995;13:577-582.

[58] Qi Q, Huang J, Crowley J, Ruschke L, Goldman BS, Wen L, Rapp WD. Metabolically engineered soybean seed with enhanced threonine levels: biochemical characterization and seed-specific expression of lysine-insensitive variants of aspartate kinases from the enteric bacterium *Xenorhabdus bovienii*. Plant Biotechnol J. 2011;9(2):193-204.

[59] Goeddel DV, Kleid DG, Bolivar F, Heyneker HL, Yansura DG, Crea R, Hirose T, Kraszewski A, Itakura K, Riggs AD. Expression in *Escherichia coli* of chemically synthesized genes for human insulin. ProcNatlAcadSci U S A. 1979;76(1):106–110.

[60] Cho MJ, Widholm JM, Vodkin LO. Cassettes for seed-specific expression tested in transformed embryogenie cultures of soybean. Plant Mol. Binl. Reporter. 1995;13:225-269.

[61] Maughan PJ, Philip R, Cho MJ, Widholm JM, Vodkin LO. Biolistic transformation, expression, and inheritance of bovine β-casein in soybean (*Glycine max*). In Vitro Cell. Devel. Biol. Plant. 1999;35:344–349.

[62] Philip R, Darnowski DW, Maughan PJ, Vodkin LO. Processing and localization of bovine β-casein expressed intransgenic soybean seeds under control of a soybean lectin expression cassette. Plant Science. 2001;161:323-333.

[63] Zeitlin L, Olmsted SS, Moench TR, Co MS, Martinell BJ, Paradkar VM, Russell DR, Queen C, Cone RA, Whaley KJ. A humanized monoclonal antibody produced in transgenic plants for immunoprotection of the vagina against genital herpes. Nature Biotechnol. 1998;16:1361–1364.

[64] Oakes, JL, Bost, KL, Piller, KJ. Stability of a soybean seed-derived vaccine antigen following long-term storage, processing and transport in the absence of a cold chain. Journal of the Science of Food and Agriculture. 2009;89(13):2191-2199.

[65] Cunha, NB, Araujo, AC, Leite, A, Murad, AM, Vianna, GR, Rech, EL. Correct targeting of proinsulin in protein storage vacuoles of transgenic soybean seeds. Genet Mol Res. 2010;9(?):1163-1170.

[66] Cunha NB, Murad AM, Cipriano TM, Cla'udia A, Araujo G,.Aragaǒ FJL, Leite A, Vianna GR, McPhee TR, Souza GHMF, Waters MJ, Elı́bio L. Rech. Expression of functional recombinant human growth hormone in transgenic soybean seeds. Transgenic Res. 2011;20:811–826.

[67] Ryan ET, Crean TI, John M, Butterton JR, Clements JD, Calderwood SB. *In vivo* expression and immunoadjuvancy of a mutant of heat-labile enterotoxin of *Escherichia coli* in vaccine and vector strains of *Vibriocholerae*. Infect Immun. 1999;67(4):1694–1701.

[68] Moravec T, Schmidt MA, Herman EM, Woodford-Thomas T. Production of *Escherichia coli* heat labile toxin (LT) B subunit in soybean seed and analysis of its immunogenicity as an oral vaccine. Vaccine. 2007;19;25(9):1647-1657.

[69] Piller KJ, Clemente TE, Jun SM, Petty CC, Sato S, Pascual DW, Bost KL. Expression and immunogenicity of an Escherichia coli K99 fimbriae subunit antigen in soybean. Planta. 2005;222(1):6-18.

[70] Vimolmangkang S, Gasic K, Soria-Guerra R, Rosales-Mendoza S, Moreno-Fierros L, Korban SS. Expression of the nucleocapsid protein of porcine reproductive and respiratory syndrome virus in soybean seed yields an immunogenic antigenic protein. Planta. 2012;235(3):513-522.

[71] Russell, DA, Spatola, LA, Dian, T, Paradkar, VM, Dufield, DR, Carroll, JA and Schlit-
      tler, MR. Host limits to accurate human growth hormone production in multiple
      plant systems. Biotechnol. Bioeng. 2005;89:775–782.

[72] Ding SH, Huang LY, Wang YD, Sun HC, Xiang ZH. High-level expression of basic
      fibroblast growth factor in transgenic soybean seeds and characterization of its bio-
      logical activity. BiotechnolLett. 2006;28(12):869-875.

[73] MatobaN ,Usui H, Fujita H, Yoshikawa M. A novel anti-hypertensive peptide de-
      rived from ovalbumin induces nitric oxide-mediated vasorelaxation in an isolated
      SHR mesenteric artery FEBS Lett. 1999;452:181–184.

[74] Yamada Y, Nishizawa K, Yokoo M, Zhao H, Onishi K, Teraishi M, Utsumi S, Ishimo-
      to M, Yoshikawa M. Anti-hypertensive activity of genetically modified soybean
      seeds accumulating novokinin. Peptides. 2008;29(3):331-337.

[75] Cunha NB, Murad AM, Ramos GL, Maranhão AQ, Brígido MM, Araújo AC, Lacorte
      C, Aragão FJ, Covas DT, Fontes AM, Souza GH, Vianna GR, Rech EL. Accumulation
      of functional recombinant human coagulation factor IX in transgenic soybean seeds.
      Transgenic Res. 2011;20(4):841-855.

[76] Powell R, Hudson LC, Lambirth KC, Luth D, Wang K, Bost KL, PillerKJ. Recombi-
      nant expression of homodimeric 660 kDa human thyroglobulin in soybean seeds: an
      alternative source of human thyroglobulin. Plant Cell Rep. 2011;30(7):1327-1338.

[77] Schmidt MA, Herman EM. Proteome rebalancing in soybean seeds can be exploited
      to enhance foreign protein accumulation. Plant Biotechnol J. 2008;6(8):832-842.

Chapter 4

Functional Diversity of Early Responsive to Dehydration (ERD) Genes in Soybean

Murilo Siqueira Alves and Luciano Gomes Fietto

Additional information is available at the end of the chapter

1. Introduction

In many regions of the world, agriculture is the primary consumer of water. As the world population increases, and arid regions become more abundant, water will become an increasingly scarce resource [1]. In 2011, the world's soybean crops produced 263.7 million tons from an area of 103,5 million hectares [2]. This global production required an input of 0,2 to 0,25 inch of water per acre per day during peak demand, which represents a major problem for the producer countries [3]. In Brazil, the second largest soybean producer in the world, there was a 7% reduction in soybean production in 2011/2012 compared to the previous season. This yield loss can be attributed to drought in the soybean-growing regions of the country, which in turn resulted in increased use of irrigation water in an attempt to minimize yield losses [4].

Understanding the molecular consequences of drought on soybean plants can accelerate breeding programs aimed at increasing productivity and decreasing the negative impacts of climate change on this important crop. Several classical physiology reviews from recent decades consolidated knowledge of the relationship between leaf structure and function during drought stress [5,6], the morphology of the root during stress tolerance [5,6] and other aspects of the effects of drought on plant morphology. Understanding the physiological responses of plants undergoing drought stress is essential to understanding their ability to survive the water shortage.

In recent years, due to advancements in plant molecular biology methodologies, molecular aspects of drought tolerance have received special attention from researchers [7]. To date, hundreds of genes that are induced by drought stress have been identified and a range of genetic, biochemical and molecular assays (gene expression profiles, transgenic plants, and various functional assays), are being used to elucidate the roles of these genes in response to drought. However, the complexity of the plant response to drought stress makes it difficult to identify genes that are responsible for drought tolerance [7]. In physiological terms, drought stress is characterized by reduction in plant water content, decrease in water potential, loss of leaf turgor, stomata closure and reduction in cell growth [8]. Conditions of severe and prolonged drought result in cessation of photosynthesis, metabolic disorder, and finally plant death [8].

Many abiotic stresses such as high salt levels and low temperature have similar physiological consequences to drought, and therefore similar signaling pathways are induced [7]. The similarity of the cold and drought stress response is illustrated by the observation that plants subjected to drought stress display an increase in frost tolerance [9]. An increase in osmotic pressure is common to these abiotic stresses [10]. The increased osmolarity induces transcription of genes encoding proteins involved in synthesis of osmo-protective compounds, lipid desaturases and transcription factors [11]. Several of these genes have been frequent targets of genetic engineering in breeding programs aimed at producing cultivars with increased tolerance to these adverse conditions [11]. These genes are also induced by other environmental factors such as high salinity and chemical signals such as abscisic acid (ABA), the main phytohormone related to abiotic stress responses in plants.

ABA serves as an endogenous messenger in response to biotic and abiotic stress in plants. Drought results in production of high levels of ABA, accompanied by a major shift in global gene expression in plant cells and, consequently, an adaptive physiological response to the stress [12]. In addition to stress, ABA also controls other important and finely regulated processes such as growth and development, structure and regulation of stomatal function and seed dormancy [13]. During regulation of plant development, ABA also acts in intricate cross-communication with other important phytohormones, such as gibberellic acid, ethylene, auxin and brassinosteroids [13].

How and what environmental stimuli are perceived and result in changes in physiological levels of ABA is still a difficult issue. Drought stress provides an immediate hydraulic signal to the plant, which activates ABA biosynthesis over a great distance [14]. High humidity activates cytochrome P450 enzymes that catalyze ABA synthesis minutes after perception of the stress [15]. Recent studies have shown the importance of the transport driven by absorption and export of ABA. Upon perception of the stress signal, ABA synthesis is primarily induced in vascular tissues, and ABA is exported from the site of biosynthesis to other cells. The absorption is stimulated by ATP-dependent ABC-family transporters. This mechanism allows rapid distribution of ABA to the surrounding tissues [16,17].

Although expression of many genes is induced by ABA-dependent responses to drought, cold and salinity stresses, upregulated genes can be sub-grouped according to the stress they were found to respond to and also by the timing of induction post stress. Genes included in the RD group (responsive to dehydration) include the drought-induced gene RD26, which encodes a NAC (NAM/ATAF/CUC plant protein domains)-family transcription factor [18], ERD (early responsive to dehydration), which includes a gene that encodes a Clp protease [19]. The COR (cold regulated), LTI (low-temperature induced) groups of genes include LOS2, which encodes a bifunctional enolase [20]. The KIN (cold inducible) group of genes includes SCOF-1, which encodes a protein with a zinc finger domain [21]. The KIN group also contains groups of genes, which also respond to osmotic stress [22-23, 7]. The products of many of these genes are most likely the main components of the first line of plant defense against potential structural damage, or they may be components of signaling pathways such as transcription factors or protein kinases. An example is induction of the gene COR15a; the Arabidopsis homolog ERD1 prevents the injury to the chloroplast membrane [24]. Another gene, GmERD15, from the ERD15 gene family in soybean, acts as a transcription factor, which regulates gene transcription related to programmed cell death [25].

2. The Early Responsive to Dehydration (ERD) genes and their functional diversity

The ERD genes are defined as those genes that are rapidly activated during drought stress. The encoded proteins show a great structural and functional diversity and constitute the first line of defense against drought stress in plants (Table 1).

To date, a total of 16 complementary DNAs (cDNAs) for ERD genes have been isolated from 1-h-dehydrated Arabidopsis thaliana and only half of these are characterized in soybean. These genes encode proteins that include ClpA/B adenosine triphosphate (ATP)-dependent protease, heat shock protein (HSP) 70-1, S-adenosine-methionine-dependent methyltransferases, membrane protein, proline dehydrogenase, sugar transporter, senescence-related gene, glutathione-S-transferase, group II LEA (Late Embryogenesis Abundant) protein, chloroplast and jasmonic acid biosynthesis protein, hydrophilic protein, and ubiquitin extension protein.

Gene / GenBank accession number	Function	Reference	Best hit on soybean genome / E value.	Similar genes in soybean
ERD1/D17582*	ClpA/B ATP-dependent protease	[26]	Glyma04g38050.1/8.9e-54	45
ERD2*/M23105	Heat shock protein (hsp70-i)	[26]	Glyma12g06910.1/ 1.7e-37	45
ERD3/NP_567575.1*	Methyltransferase PMT21	[27]	Glyma01g35220.4/ 1.7e-99	100
ERD4/NP_564354.1	Integral membrane protein	[28]	Glyma15g09820.1/ 1.8e-61	28
ERD5/D83025	Precursor of proline dehydrogenase	[29]	Glyma18g51400.1/ 2.1e-27	5
ERD6/D89051	Sugar transporter	[30]	Glyma03g40160.1/ 6.15e-2	100

Gene / GenBank accession number	Function	Reference	Best hit on soybean genome / E value.	Similar genes in soybean
ERD7/NP_179374.1	Senescence/ dehydration related protein	[31]	Glyma01g36960.1/ 3.5e-43	9
ERD8/Y11827	Heat shock protein hsp81-2)	[26]	Glyma08g44590.1/0	22
ERD9/NP_172508.4*	Glutathione-S-transferase	[32]	Glyma01g04710.1/ 9.9e-22	87
ERD10/D17714*	Group II LEA protein (lti29/lti45)	[33]	Glyma04g01130.1/ 4e-8	3
ERD11/D17672	Glutathione-S-transferase	[32]	Glyma02g17340.1/ 1.7e-5	52
ERD12/NP_189204.1*	Allene oxide cyclase	[40]	Glyma02g11020.1/ 2.3e-36	6
ERD13/D17673	Glutathione S-transferase	[32]	Glyma08g41960.1/ 3.7e-39	63
ERD14/D17715	Group II LEA protein	[33]	No homologs identified	0
ERD15/D30719*	Hydrophilic protein	[35]	Glyma04g28560.1/ 3.5e-26	4
ERD16/J05507*	Ubiquitin extension protein	[33]	Glyma03g35540.1/ 6.6e-50	100

Table 1. ERD genes and their homologs in soybean. (*) Indicates the characterized genes in soybean.

The ERD gene family has been collectively characterized as genes that are rapidly induced by dehydration [26]. ERD1 encodes a chloroplast ATP-dependent protease [26] and ERD2 encodes a, HSP70 [26], ERD3 encodes a methyltransferase in the pMT21 family [27], ERD4 encodes a membrane protein [28], ERD5 and ERD6 encode a mitochondrial dehydrogenase proline protein and a carbohydrates carrier protein, respectively [29-30]. ERD7 encodes a protein related to senescence and dehydration [31], ERD8 encodes a hsp81-family protein [26], ERD9, 11 and 13 belong to the family of glutathione S-transferase [32], ERD10 and 14 belong to the LEA protein family [33], ERD15 was first classified as a hydrophilic protein [34], which has a PAM2 interaction domain which interacts with poly-A tail binding proteins (PABP) [35].

ERD15 from Arabidopsis has been functionally characterized as a common regulator of the abscisic acid (ABA) response and salicylic acid (SA)-dependent defense pathway [35]. Overexpression of ERD15 reduced ABA sensitivity, as the transgenic plants had reduced drought tolerance and failed to increase their freezing tolerance in response to hormone treatment [35]. In contrast, loss of ERD15 function due to gene silencing caused hypersensitivity to ABA, and the silenced plants displayed enhanced tolerance to both drought and freezing. The antagonistic effect of ERD15 activity on ABA signaling enhanced SA-dependent defense; overexpression of ERD15 was associated with increased resistance to the bacterial necrotroph Erwinia carotovora and enhanced induction of systemic acquired resistance reporter genes [35]. The authors also addressed the antagonistic effect of ABA on SA-mediated defense by demonstrating the enhanced expression of reporter genes for systemic acquired resistance in the plant null mutants abi1-1 and abi2-1, which are defective for ABA metabolism. These results together implicate Arabidopsis ERD15 as a shared component of ABA- and SA-mediated responses. The ERD15 homologs from Solanum licopersicum are 98% identical and belong to the same group as Arabidopsis ERD15, indicating a possible conservation of function [36]. Nevertheless, the tomato protein clearly localizes to the nucleus and confers freezing tolerance when ectopically expressed in transgenic tomato plants. These phenotypes are in marked contrast with the phenotypes displaying by ERD15-overexpressing Arabidopsis lines [35]. These contrasting results in transgene overexpression studies suggest that the Arabidopsis and tomato ERD15 homologs have divergent functions. Finally, a soybean homolog, GmERD15, has been described as an ER stress- and osmotic stress-induced transcription factor that activates the promoter and induces the expression of the NRP-B gene. These results indicate that GmERD15 functions as an upstream component of the NRP-mediated cell death signaling pathway, which is induced by ER and osmotic stress [37].

ERD16 encodes a ubiquitination extension protein [33]. Previous studies also showed that ERD13/AtGSTF10, a plant phi specific class GST (Glutathione S-transferase) is an interaction protein with BAK1 (BRI1 Associated receptor Kinase 1). BAK1 is a co-receptor, which forms a receptor complex with BRI1 (brassinosteroid (BR) receptor) to regulate brassinosteroid signaling in Arabidopsis. Overexpression of AtGSTF10 resulted in plants with increased tolerance to salt stress. In contrast, silencing AtGSTF10 by RNAi caused increased tolerance to abiotic stress and accelerated senescence of the transformants [38]. These findings suggest that modulation of ERD13/AtGSTF10 may regulate plant stress responses by regulating brassinosteroid signaling via interaction of AtGSTF10 with BAK1. ERD10 and 14 have chaperone activity, which aid in protein folding during stress [39]. ERD12 encodes a protein with homology to an allene oxide cyclase [40].

With respect to expression controlled by phytohormones, ERD genes present varied functions and responses in ABA signaling, some being sensitive to ABA during germination and development [41], and /or are involved in stress tolerance [42]. Other genes are induced in response to more than one phytohormone [35]. Early Responsive to Dehydration 15 (ERD15) was characterized as a negative regulator of ABA and is induced by ABA, SA, injury and pathogen infection [35]. ABA application increases the expression of some members of the ERD group including ERD10 and 14 [34] while causing no effect on others, such as ERD2, 8 and 16 [33].

Some contradictory data regarding the induction, as well as the function, of ERD genes are present in the literature [35-37]. Reduced expression of the ERD15 gene in response to wounding was reported [43], while an increased number of ERD15 transcripts were observed by other authors [35]. Furthermore, Arabidopsis plants showed increased tolerance to salt stress through the overexpression of AtSAT32, a key gene in the salinity-tolerance family Arabidopsis. These plants showed an increase in the number of ERD15 transcripts relative to control plants [44]. Transgenic wheat plants over-expressing TaDi19A, a gene responsive to salinity in wheat, exhibited increased expression of ERD15 [45]. In contrast to these findings, Arabidopsis plants over-expressing ERD15 demonstrated susceptibility to drought and freezing [35]. In regard to function, a soybean ERD15 homolog was characterized as a transcription factor [25], a function not previously attributed to this protein family, as reported by Kariola and colleagues [35] and Ziaf and colleagues [36].

3. The ERD genes studied in soybean

In respect to ERD genes described in soybean, the behavior of a group of eight genes (ERD1, ERD2, ERD3, ERD9, ERD10, ERD12, ERD15 gene and ERD16) was studied in response to stress. A soybean cDNA ERD1, homologous to yeast Hsp104, was isolated and characterized [46]. The soybean genes encoding homologs to yeast Hsp104 and Hsp101 have a high level of sequence identity to members of the family Clp [46]. When heterologously expressed in yeast, the soybean Hsp101 gene conferred greater thermotolerance to yeast [46]. Several genes related to Hsp70 (ERD2) have been described in soybean using proteomics studies. The first evidence of an ERD2-like protein in soybean was found during heat shock [47]. The presence of similar proteins was also found in response to osmotic and reticulum stress [48]. In respect to the orthologs of ERD3 in soybean, it was found that the GmIMT gene, which encodes a methyltransferase, acts by methylating the substrate D-ononitol. Its overexpression in Arabidopsis causes an increase in drought and salinity tolerance [49]. When a gene encoding a soybean GST, an ortholog of ERD9, was over-expressed in tobacco plants, it conferred an increase in salinity and drought tolerance [50].

Group 2 LEA (dehydrins or responsive to abscisic acid) proteins, such as ERD10 proteins, are postulated to protect macromolecules from damage by freezing, dehydration, ionic, or osmotic stress. In soybean, proteins of this group were studied for their structural and physio-chemical properties but little is discussed regarding the function of these proteins [51]. Overexpression of a member of the ERD12 family, GmAOC5, significantly increased oxidative stress resistance [52]. Within the ERD15 family, a soybean ortholog, GmERD15 has been functionally characterized as a transcription factor; in response to osmotic stress, GmERD15 acts to control transcription of a gene related to an integrative pathway in soybean [25]. Finally, orthologs of ERD16 studied in soybean genes were identified with differential expression during flood stress and hypoxia [53]. All genes studied related to different levels in response to stress, particularly drought and osmotic stress, demonstrating the conservation of function of this gene family in different plant species.

Some ERD genes not yet studied in soybean deserve special attention because of either the proven involvement of a gene with similar functions in drought response in other organisms, or due to multiple copies of the soybean homolog. ERD5 and ERD7 family members have been characterized by activity in response to drought in other organisms (discussed below). They have not been studied in soybean; however, homology to soybean genes is demonstrated by the phylogenic tree shown in Figure 1.

ERD5 (which encodes a precursor of a proline dehydrogenase), has five orthologous genes in soybean. ERD5 has a proven role in drought response due to its role in accumulation of proline [54], a common occurrence during osmotic stress. All soybean genes are clustered in a group distinct from orthologs in other species (Figure 1), which may reflect a possible functional divergence.

ERD7 (which encodes a protein related to senescence and dehydration) has nine orthologous genes in soybean. It also has a central role in response to drought and osmotic stress and it is related with drought-induced leaf senescence in plants [55]; regulation of this process during drought tolerance has been studied in depth [55]. Phylogenetic analysis of these genes suggests the possibility of functional divergence of these genes within the same organism.

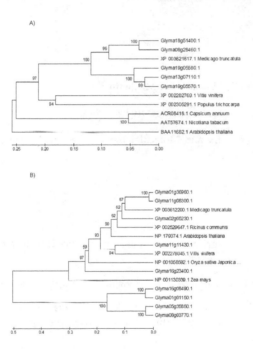

Figure 1. Relatedness of ERD5 (panel A) and ERD7 (panel B) proteins from different plant species. The multiple alignment was made using ClustalW, and the dendrogram was built with the MEGA5 software using the UPGMA method. The numbers at the nodes indicate the bootstrap scores. The proteins accession numbers are indicated.

4. Conclusion

Many studies on the roles and importance of ERD genes in soybean have become necessary due to lack of information about the importance of this group of genes during plant response to drought. The common feature of these genes is that their expression increases rapidly in response to drought stress, suggesting that it is the first line of defense for plants against drought stress. It also suggests these genes may function to regulate expression of effector proteins and signaling pathways in response to stress.

Acknowledgements

This research was supported by the Conselho Nacional de Desenvolvimento Científico e Tecnológico CNPq and Fundação de Amparo a Pesquisa do Estado de Minas Gerais (FAPE-MIG). M.S.A. is supported by fellowship from CAPES.

Author details

Murilo Siqueira Alves and Luciano Gomes Fietto*

*Address all correspondence to: lgfietto@ufv.br

Department of Biochemistry and Molecular Biology, Federal University of Viçosa, Viçosa, Minas Gerais, Brazil

References

[1] FAO: Food and agriculture organization of the United Nations. (2012). http://www.fao.org/., (accessed 12 April).

[2] Embrapa Soybean. (2012). http://www.cnpso.embrapa.br, (accessed 12April).

[3] Mississippi Agricultural and Forestry Experiment Station, Mississippi State University Extension Service. (2012). http://msucares.com/pubs/publications/2185 htm., (accessed 12 April).

[4] Brazilian Institute of Geography and Statistics- IBGE. (2012). www.ibge.gov.br/., (accessed 12 April).

[5] Smirnoff, N. (1998). Plant resistance to environmental stress. *Current opinion in Biotechnology*, 9-214.

[6] Gorantla, M., Babu, P. R., Lachagari, V. B. R., Reddy, A. M. M., Wusirika, R., Ennetzen, J. L., & Reddy, A. R. (2007). Identification of stress-responsive genes in anindica rice (Oryza sativa L.) using ESTs generated from drought-stressed seedlings. *Journal of Experimental Botany*, 58(2), 253-265.

[7] Shinozaki, K., & Yamaguchi-Shinozaki, K. (2007). Gene networks involved in drought stress response and tolerance. *Journal of Experimental Botany*, 58-221.

[8] Jaleel, C. A., Manivannan, P., Lakshmanan, G. M. A., Gomathinayagam, M., & Panneerselvam, R. (2008). Alterations in morphological parameters and photosynthetic pigment responses of Catharanthus roseus under soil water deficits. *Colloids and Surfaces B: Biointerfaces*, 61-298.

[9] Guy, C. L., Huber, J. L. A., & Huber, S. C. (1992). Sucrose Phosphate Synthase and Sucrose Accumulation at Low Temperature. *Plant Physiology*, 100-502.

[10] Munns, R., & Tester, M. (2008). Mechanisms of salinity tolerance. *Annual Review in Plant Biology*, 59-651.

[11] Yamagushi-Shinosaki, K., & Shinosaki, K. (2005). Organization of cis-acting regulatory elements in osmotic- and cold-stressresponsive promoters. *TRENDS in Plant Science*, 10(2), 88-94.

[12] Christmann, A., Moes, D., Himmelbach, A., Yang, Y., Tang, Y., & Grill, E. (2006). Integration of abscisic acid signalling into plant responses. *Plant Biology*, 8-314.

[13] Hirayama, T., & Shinozaki, K. (2007). Perception and transduction of abscisic acid signals: keys to the function of the versatile plant hormone ABA. *Trends of Plant Science*, 12-343.

[14] Christmann, A., Weiler, E.W., Steudle, E., & Grill, E. (2007). A hydraulic signal in root-to-shoot signalling of water shortage. *Plant Journal*, 52, 167-174.

[15] Okamoto, M., Tanaka, Y., Abrams, S. R., Kamiya, Y., Seki, M., & Nambara, E. (2009). High humidity induces abscisic acid 80 hydroxylase in stomata and vasculature to regulate local and systemic abscisic acid responses. *Arabidopsis. Plant Physiology*.

[16] Kang, J., Hwang, J.U., Lee, M., Kim, Y.Y., Assmann, S.M., Martinoia, E., & Lee, Y. (2010). PDR-type ABC transporter mediates cellular uptake of the phytohormone abscisic acid. *Proceedings of the National Academy of Sciences USA*, 107, 2355-2360.

[17] Kuromori, T., Miyaji, T., Yabuuchi, H., Shimizu, H., Sugimoto, E., Kamiya, A., Moriyama, Y., & Shinozaki, K. (2010). ABC transporter AtABCG25 is involved in abscisic acid transport and responses. *Proceedings of the National Academy of Sciences USA*, 107-2361.

[18] Fujita, M., Fujita, Y., Maruyama, K., Seki, M., Hiratsu, K., Ohme-Takagi, M., Tran, L. S., Yamaguchi-Shinozaki, K., & Shinozaki, K. (2004). A dehydration-induced NAC protein, RD26, is involved in a novel ABA-dependent stress-signaling pathway. *Plant Journal*, 39(6), 863-876.

[19] Nakashima, K., Kiyosue, T., Yamaguchi-Shinozaki, K., & Shinozaki, K. (1997). A nuclear gene, erd1, encoding a chloroplast-targeted Clp protease regulatory subunit ho-

molog is not only induced by water stress but also developmentally up-regulated during senescence in Arabidopsis thaliana. *Plant Journal*, 12(4), 851-861.

[20] Lee, H., Guo, Y., Ohta, M., Xiong, L., Stevenson, B., & Zhu, J. K. (2002). LOS2, a genetic locus required for cold-responsive gene transcription encodes a bi-functional enolase. *EMBO Journal*, 21(11), 2692-2702.

[21] Kim, J. C., Lee, S. H., Cheong, Y. H., Yoo, C. M., Lee, S. I., Chun, H. J., Yun, D. J., Hong, J. C., Lee, S. Y., Lim, C. O., & Cho, M. J. (2001). A novel cold-inducible zinc finger protein from soybean, SCOF-1, enhances cold tolerance in transgenic plants. *Plant Journal*, 25(3), 247-259.

[22] Ingram, J., & Bartels, D. (1996). The molecular basis of dehydration tolerance in plants. *Annual Reviews in Plant Physiology and Plant Molecular Biology*, 47-377.

[23] Thomashow, M.F. (1999). Plant cold acclimation: freezing tolerance genes and regulatory mechanisms. *Annual Reviews in Plant Physiology and Plant Molecular Biology*, 50-571.

[24] Steponkus, P. L., Uemura, M., Joseph, R. A., Gilmour, S. J., & Thomashow, M. F. (1998). Mode of action of the COR15a gene on the freezing tolerance of Arabidopsis thaliana. *Proceedings of the National Academy of Sciences USA*, 95-14570.

[25] Alves, M. S., Reis, P. A. B., Dadalto, S. P., Faria, J. A. Q. A., Fontes, E. P. B., & Fietto, L. G. (2011). A Novel Transcription Factor, ERD15 (Early Responsive to Dehydration 15), Connects Endoplasmic Reticulum Stress with an Osmotic Stress-induced Cell Death Signal. *Journal of Biological Chemistry*, 286(22), 20020-20030.

[26] Kiyosue, T., Yamaguchi-Shinozaki, K., & Shinozaki, K. (1994). Cloning of cDNAs for genes that are early-responsive to dehydration stress (ERDs) in Arabidopsis thaliana L.: identification of three ERDs as HSP cognate genes. *Plant Molecular Biology*, 25-791.

[27] González-Martínez, S. C., Ersoz, E., Brown, G. R., Wheeler, N. C., & Neale, D. B. (2006). DNA sequence variation and selection of tag single-nucleotide polymorphisms at candidate genes for drought-stress response. *Pinus taeda L. Genetics*, 172(3), 1915-1926.

[28] Froehlich, J. E., Wilkerson, C. G., Ray, W. K., Mc Andrew, R. S., Osteryoung, K. W., Gage, D. A., & Phinney, B. S. (2003). Proteomic study of the Arabidopsis thaliana chloroplastic envelope membrane utilizing alternatives to traditional two-dimensional electrophoresis. *Journal of Proteome Research*, 2(4), 413-425.

[29] Nakashima, K., Satoh, R., Kiyosue, T., Yamaguchi-Shinozaki, K., & Shinozaki, K. (1998). A gene encoding proline dehydrogenase is not only induced by proline and hypoosmolarity, but is also developmentally regulated in the reproductive organs of Arabidopsis. *Plant Physiology*, 118-1233.

[30] Kiyosue, T., Abe, H., Yamaguchi-Shinozaki, K., & Shinozaki, K. (1998). ERD6, a cDNA clone for an early dehydration-induced gene of Arabidopsis, encodes a putative sugar transporter. *Biochimica Biophysica Acta*, 1370-187.

[31] Kimura, M., Yamamoto, Y. Y., Seki, M., Sakurai, T., Sato, M., Abe, T., Yoshida, S., Manabe, K., Shinozaki, K., & Matsui, M. (2003). Identification of Arabidopsis genes regulated by high light-stress using cDNA microarray. *Photochemical and Photobiology*, 77(2), 226-33.

[32] Kiyosue, T., Yamaguchi-Shinozaki, K., & Shinozaki, K. (1993). Characterization of two cDNAs (ERD11 and ERD13) for dehydration inducible genes that encode putative glutathione S-transferases. *Arabidopsis thaliana L. FEBS Letters*, 335-189.

[33] Kiyosue, T., Yamaguchi-Shinozaki, K., & Shinozaki, K. (1994). Characterization of two cDNAs (ERD10 and ERD14) corresponding to genes that respond rapidly to dehydration stress. *Arabidopsis thaliana. Plant Cell Physiology*, 35-225.

[34] Kiyosue, T., Yamaguchi-Shinozaki, K., & Shinozaki, K. (1994). ERD15, a cDNA for a dehydration-inducible gene from Arabidopsis thaliana. *Plant Physiology*, 106-1707.

[35] Kariola, T., Brader, G., Helenius, E., Li, J., Heino, P., & Palva, E. T. (2006). EARLY RESPONSIVE to DEHYDRATION 15, a negative regulator of abscisic acid responses in Arabidopsis. *Plant Physiology*, 142-1559.

[36] Ziaf, K., Loukehaich, R., Gong, P., Liu, H., Han, Q., Wang, T., Li, H., & Ye, Z. (2011). A multiple stress responsive gene ERD15 from Solanum pennellii confers stress tolerance in tobacco. *Plant Cell Physiology*, 52(6), 1055-1067.

[37] Alves, M. S., Fontes, E. P. B., & Fietto, L. G. (2011). EARLY RESPONSIVE to DEHYDRATION 15, a new transcription factor that integrates stress signaling pathways. *Plant signaling & behavior*, 6(12).

[38] Ryu, H. Y., Kim, S. Y., Park, H. M., You, J. Y., Kim, B. H., Lee, J. S., & Nam, K. H. (2009). Modulations of AtGSTF10 expression induce stress tolerance and BAK1 mediated cell death. *Biochemical and Biophysical Research Communications*, 379, 417-422.

[39] Kovacs, D., Kalmar, E., Torok, Z., & Tompa, P. (2008). Chaperon activity of ERD10 and ERD14, two disordered stress-related plant proteins. *Plant Physiology*, 147-381.

[40] Gutiérrez, R. A., Green, P. J., Keegstra, K., & Ohlrogge, J. B. (2004). Phylogenetic profiling of the Arabidopsis thaliana proteome: what proteins distinguish plants from other organisms? *Genome Biology*, 5(8), R53, doi:10.1186/gb-2004-5-8-r53.

[41] Zhang, X. L., Zhang, Z. J., Chen, J., Chen, Q., Wang, X. C., & Huang, R. F. (2005). Expressing TERF1 in tobacco enhances drought tolerance and abscisic acid sensitivity during seedling development. *Planta*, 222-494.

[42] Xiong, L., Lee, B. H., Ishitani, M., Lee, H., Zhang, C., & Zhu, J. K. (2001). FIERY1 encoding an inositol polyphosphate 1 phosphatase is a negative regulator of abscisic acid and stress signaling. *Arabidopsis. Genes Development, 15,* 1971-1984.

[43] Dunaeva, M., & Adamska, I. (2001). Identification of genes expressed in response to light stress in leaves of Arabidopsis thaliana using RNA differential display. *European Journal of Biochemistry,* 268-5521.

[44] Park, M. Y., Chung, M. S., Koh, H. S., Lee, D. J., Ahn, S. J., & Kim, C. S. (2009). Isolation and functional characterization of the Arabidopsis salt-tolerance 32 (AtSAT32) gene associated with salt tolerance and ABA signaling. *Physiologia Plantarum,* 135-426.

[45] Li, S., Xu, C., Yang, Y., & Xia, G. (2010). Functional analysis of TaDi19A, a salt responsive gene in wheat. *Plant Cell Environment, 33-117.*

[46] Lee, Y. R., Nagao, R. T., & Key, J. L. (1994). A soybean 101-kD heat shock protein complements a yeast HSP104 deletion mutant in acquiring thermotolerance. *Plant Cell, 6*(12), 1889-1897.

[47] Roberts, J. K., & Key, J. L. (1991). Isolation and characterization of a soybean Hsp70 gene. *Plant Molecular Biology, 1*(6), 671-683.

[48] Valente, M. A., Faria, J. A., Soares-Ramos, J. R., Reis, P. A., Pinheiro, G. L., Piovesan, N. D., Morais, A. T., Menezes, C. C., Cano, M. A., Fietto, L. G., Loureiro, M. E., Aragão, F. J., & Fontes, E. P. (2009). The ER luminal binding protein (BiP) mediates an increase in drought tolerance in soybean and delays drought-induced leaf senescence in soybean and tobacco. *Journal of Experimental Botany, 60*(2), 533-546.

[49] Ahn, C., Park, U., & Park, P. B. (2011). Increased salt and drought tolerance by D-ononitol production in transgenic Arabidopsis thaliana. *Biochemical and Biophysical Research Communications, 415*(4), 669-674.

[50] Ji, W., Zhu, Y., Li, Y., Yang, L., Zhao, X., Cai, H., & Bai, X. (2010). Over-expression of a glutathione S-transferase gene, GsGST, from wild soybean (Glycine soja) enhances drought and salt tolerance in transgenic tobacco. *Biotechnology Letters, 32*(8), 1173-1179.

[51] Soulages, J. L., Kim, K., Arrese, E. L., Walters, C., & Cushman, J. C. (2003). Conformation of a group 2 late embryogenesis abundant protein from soybean. Evidence of poly(L-proline)-type II structure. *Plant Physiology, 131*(3), 963-975.

[52] Wu, Q., Wu, J., Sun, H., Zhang, D., & Yu, D. (2011). Sequence and expression divergence of the AOC gene family in soybean: insights into functional diversity for stress responses. *Biotechnology Letters, 33*(7), 1351-1359.

[53] Yanagawa, Y., & Komatsu, S. (2012). Ubiquitin/proteasome-mediated proteolysis is involved in the response to flooding stress in soybean roots, independent of oxygen limitation. *Plant Science.*

[54] Verbruggen, N., & Hermans, C. (2008). Proline accumulation. *Plants: a review. Amino Acids*, 35(4), 753-759.

[55] Munné-Bosch, S., & Alegre, L. (2004). Die and let live: leaf senescence contributes to plant survival under drought stress. *Functional Plant Biology*, 31-203.

Chapter 5

Gene Duplication and RNA Silencing in Soybean

Megumi Kasai, Mayumi Tsuchiya and
Akira Kanazawa

Additional information is available at the end of the chapter

1. Introduction

Soybean, *Glycine max* (L.) Merr., is considered to be a typical paleopolyploid species with a complex genome [1-3]. Approximately 70 to 80% of angiosperm species have undergone polyploidization at some point in their evolutionary history, which is a well-known mechanism of gene duplication in plants [4]. The soybean genome actually possesses a high level of duplicate sequences, and furthermore, possesses homoeologous duplicated regions, which are scattered across different linkage groups [5-8]. Based on the genetic distances estimated by synonymous substitution measurements for the pairs of duplicated transcripts from expressed sequence tag (EST) collections of soybean and *Medicago truncatula*, Schlueter et al. estimated that soybean probably underwent two major genome duplication events: one that took place 15 million years ago (MYA) and another 44 MYA [9].

Gene duplication is a major source of evolutionary novelties and can occur through duplication of individual genes, chromosomal segments, or entire genomes (polyploidization). Under the classic model of duplicate gene evolution, one of the duplicated genes is free to accumulate mutations, which results in either the inactivation of transcription and/or a function (pseudogenization or nonfunctionalization) or the gain of a new function (neofunctionalization) as long as another copy retains the requisite physiological functions [10; and references therein]. However, empirical data suggest that a much greater proportion of gene duplicates is preserved than predicted by the classic model [11].

Recent advances in genome study have led to the formulation of several evolutionary models: a model proposed by Hughes [12] suggests that gene sharing, whereby a single gene encodes a protein with two distinct functions, precedes the evolution of two functionally distinct proteins; the duplication–degeneration–complementation model suggests that duplicate genes acquire debilitating yet complementary mutations that alter one or more subfunctions of the single gene progenitor, an evolutionary consequence for duplicated loci referred to as subfunctionalization [4, 11, 13]. In addition to this notion, models involving epigenetic silencing of duplicate genes [14] or purifying selection for gene balance [15, 16] have also been proposed. In soybean, differential patterns of expression have often been detected between homoeologous genes [17, 18], which indicates that subfunctionalization has occurred in these genes.

When the extent of subfunctionalization is limited, mutations in only one of multiple cognate gene copies do not often result in phenotypic changes. Therefore, methods that allow suppression of all copies of the duplicated gene are required for analyzing gene function or engineering novel traits. RNA silencing refers collectively to diverse RNA-mediated pathways of nucleotide-sequence-specific inhibition of gene expression, either at the posttranscriptional or transcriptional level, which provides a powerful tool to downregulate a gene or a gene family [19, 20]. Suppression of gene expression through RNA silencing is particularly useful for analyzing the function(s) of duplicated genes or engineering novel traits because it allows silencing of multiple cognate genes having nucleotide sequence identity. In fact, to produce soybean lines that have a novel trait, researchers have frequently used RNA silencing induced by a transgene.

In this review, we describe application of RNA silencing to understand the roles of genes or engineering novel traits in soybean. We describe methods to induce simultaneous silencing of duplicated genes and selective silencing of each copy of duplicated genes through RNA silencing. In addition to intentionally induced RNA silencing, we also refer to naturally occurring RNA silencing. Based on our knowledge of RNA silencing in soybean, we propose a hypothesis that plants may have used subfunctionalization of duplicated genes as a means to avoid the occurrence of simultaneous silencing of duplicated genes, which could be deleterious to the organism.

2. Mechanisms and diverse pathways of RNA silencing

Gene silencing is one of the regulatory mechanisms of gene expression in eukaryotes, which refers to diverse RNA-guided sequence-specific inhibition of gene expression, either at the posttranscriptional or transcriptional level [19, 20]. Post-transcriptional gene silencing (PTGS) was first discovered in transgenic petunia plants whose flower color pattern was changed as a consequence of overexpression of a gene that encodes the key enzyme for anthocyanin biosynthesis in 1990 [21, 22]. Similar phenomena have also been reported for plants transformed with various genes, which include virus resistance of plants that have gene or gene segments derived from the viral genome [23, 24]. Because of these findings, gene silencing is thought to have developed to defend against viruses. Several lines of research in plants indicated that double-stranded RNA (dsRNA) is crucial for RNA degradation [25, 26]. The potency of dsRNA to induce gene silencing was demonstrated in *Caenorhabditis elegans* by injecting dsRNA into cells in 1998 [27], and the phenomenon was termed RNA interference (RNAi).

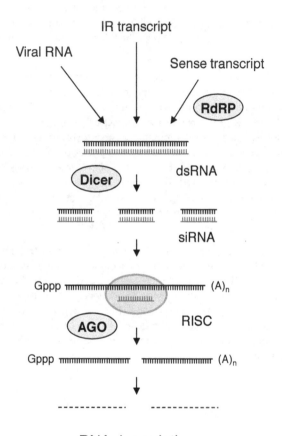

Figure 1. Pathways of RNA silencing used to downregulate a target gene through RNA degradation. Posttranscriptional gene silencing is triggered by dsRNA. Transcripts from transgenes that have an IR sequence can form dsRNA. Sense transcripts can produce dsRNA through the synthesis of complementary strand by RdRP. The replication intermediate or duplex structures formed within single-stranded RNA of the viral genome can also provide dsRNA. These dsRNAs are processed into siRNAs by the endonuclease Dicer. The siRNA is loaded into the RISC complex that contains AGO and guides the RISC complex to the mRNA by base-pairing. The RISC complex cuts the mRNA, which is subsequently degraded. Abbreviations: IR, inverted repeat; RdRP, RNA-dependent RNA polymerase; dsRNA, double-stranded RNA; siRNA, short interfering RNA; RISC, RNA-induced silencing complex; AGO, Argonaute.

Subsequent genetic and biochemical analyses in several organisms revealed that PTGS and RNAi share the same pathway and consist of two main processes: (i) processing of dsRNA into 20–26-nt small RNA molecules (short interfering RNA; siRNA) by an enzyme called Dicer that has RNaseIII-like endonuclease activity; (ii) cleavage of RNA guided by siRNA at a complementary nucleotide sequence in the RNA-induced silencing complex (RISC) containing the Argonaute (AGO) protein (Figure 1) [28]. The formation of dsRNA from single-stranded sense RNA was explained by the synthesis of its complementary strand by RNA-dependent RNA polymerase (RdRP). This process provides templates for Dicer cleavage that produces siRNAs and consequently allows amplification of silencing [29]. siRNA is responsible for not only induction of sequence-specific RNA degradation but also epigenetic changes involving DNA methylation and histone modification in the nucleus, which leads to transcriptional gene silencing (TGS) [30]. It has become evident that siRNA plays a role in systemic silencing as a mobile signal [31, 32]. In addition to siRNA, small RNA molecules called micro RNAs (miRNAs) are also involved in negative regulation of gene expression [33]. These gene silencing phenomena that are induced by sequence-specific RNA interaction are collectively called RNA silencing [34, 35].

RNA silencing plays an important role in many biological processes including development, stability of the genome, and defense against invading nucleic acids such as transgenes and viruses [20, 29, 30]. It can also be used as a tool for analyzing specific gene functions and producing new features in organisms including plants [36-38].

3. Methods of the induction of RNA silencing in soybean

3.1. Transgene-induced RNA silencing

Engineering novel traits through RNA silencing in soybean has been done using transgenes or virus vectors (Figure 1). RNA silencing in some transgenic soybean lines was induced by introducing a transgene that transcribes sense RNA homologous to a gene present in the plant genome, a phenomenon termed co-suppression [21]. This type of silencing was first discovered in transgenic petunia plants that had silencing of CHS-A for chalcone synthase [21, 22], in which mRNA transcribed from both CHS-A transgene and endogenous CHS-A gene was degraded. When sense transcripts from a transgene trigger RNA degradation, the pathway is also referred to as sense (S)-PTGS [19]. To obtain plants that have RNA silencing of a particular gene target, it is possible to generate co-suppressed plant lines as a byproduct of a transformation to overexpress the gene under the control of a strong promoter. However, a more promising method to induce RNA degradation is to transform plants with a construct comprising an inverted repeat (IR) sequence of the target gene, which forms dsRNA upon transcription (IR-PTGS) [39, 40]. This idea was based on the understanding of general mechanisms of RNA silencing in which dsRNA triggers the reaction of RNA degradation. The majority of transgene-induced RNA silencing in soybean have actually been done using such an IR construct. IR-PTGS can also be induced when multiple transgenes are integrated in the same site in the genome in an inverted orientation and fortuitous read-through transcription over the transgenes produces dsRNA.

An interesting finding reported in soybean is that RNA silencing is induced by a transgene that transcribes inverted repeats of a fatty acid desaturase *FAD2-1A* intron [41]. This result is contrary to the earlier belief that RNA silencing is a cytoplasmic event and intron does not trigger RNA degradation, which has been shown, for example, by using viral vector in plants [42] or by dsRNA injection to *C. elegans* cells [27], although irregular nuclear process-ing of primary transcripts associated with PTGS/RNAi has been reported previously [43]. The *FAD2-A1* intron-induced RNA silencing led to the understanding that RNA degrada-tion can take place in the nucleus [44]. Although whether RNA degradation in the nucleus is inducible for other genes or in other plants has not been known, this phenomenon is intrigu-ing because the involvement of nuclear events has been assumed for amplification of RNA silencing via transitivity [45] or intron-mediated suppression of RNA silencing [46].

Transcribing a transgene with a strong promoter tends to induce RNA silencing more fre-quently than that with a weak promoter [47]. For obtaining a higher level of transcription in soybean plants, the *Cauliflower mosaic virus* (CaMV) promoter has been used as in other plant species. Seed-specific promoters, such as those derived from the genes encoding subunits of β-conglycinin, glycinin, or Kunitz trypsin inhibitor, have also been used in soybean to in-duce seed-specific silencing, one feature that is exploited for metabolic engineering in soybean.

A gene construct that induces RNA silencing has been introduced to the soybean genome using either *Agrobacterium tumefaciens* infection or particle bombardment, which can pro-duce stable transgenic soybean lines that have altered traits. In addition, RNA silencing can be induced in soybean roots using *A. rhizogenes*-mediated transformation, which has been used for gene functional analysis. Methods for soybean transformation have been reviewed elsewhere [48].

3.2. Virus-induced gene silencing (VIGS)

RNA silencing has also been induced using a virus vector in soybean. Plants intrinsically have the ability to cope with viruses through the mechanisms of RNA silencing. When plants are infected with an RNA virus, dsRNA of the viral genome is degraded by the infect-ed plants [49, 50]. The dsRNA in the virus-infected cells is thought to be the replication in-termediate of the viral RNA [51] or a duplex structure formed within single-stranded viral RNA [52]. The viral genomic RNA can be processed into siRNAs, then targeted by the siR-NA/RNase complex. In this scenario, if a nonviral segment is inserted in the viral genome, siRNAs would also be produced from the segment. Therefore, if the insert corresponds to a sequence of the gene encoded in the host plant, infection by the virus results in the produc-tion of siRNAs corresponding to the plant gene and subsequently induces loss of function of the gene product (Figure 2). This fact led to the use of a virus vector as a source to induce silencing of a specific gene in the plant genome, which is referred to as virus-induced gene silencing (VIGS) [42, 53, 54]. So far, at least 11 RNA viruses and five DNA viruses were de-veloped as a plant virus vector for gene silencing, as listed previously [37]. Three vectors are now available in soybean: those based on *Bean pod mottle virus* (BPMV) [55], *Cucumber mosaic virus* (CMV) [56], and *Apple latent spherical virus* (ALSV) [57].

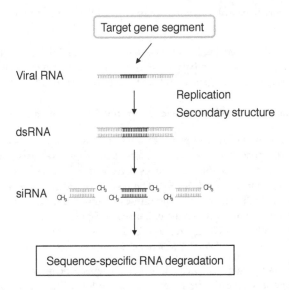

Figure 2. Virus-induced silencing of plant endogenous gene. When plants are infected with an RNA virus, dsRNA of the viral genome is degraded by the infected plants. The dsRNA in the virus-infected cells is thought to be the replication intermediate or secondary-structured viral RNA. The viral genomic RNA can be processed into siRNAs. If a plant gene segment is inserted in the viral genome, siRNAs corresponding to the plant gene are produced and subsequently induce sequence-specific RNA degradation of the plant gene.

4. Examples of RNA silencing reported in soybean

4.1. Metabolic engineering by transgene-induced RNA silencing

To the authors' knowledge, 28 scientific papers that describe metabolic engineering by transgene-induced RNA silencing in soybean have been published up to 2011 [58]. Because soybean seeds are valued economically for food and oil production, most modifications to transgenic soybean plants using RNA silencing are focused on seed components. Metabolic pathways in developing seeds have been targeted in terms of altering nutritional value for human or animals, e.g., changing seed storage protein composition [59, 60], reducing phytic acids [61, 62], saponin [63] or allergens [64], and increasing isoflavone [65]. Metabolic engineering has also targeted oil production [66-72]. These modifications were done by inhibiting a step in a metabolic pathway to decrease a product or by blocking a competing branch pathway to increase a product.

RNA silencing can be induced efficiently in soybean roots using *A. rhizogenes*-mediated root transformation. This method has been used for analyzing roles of gene products in nodule development and/or function, which occurs as a consequence of interaction between legume plants and the nitrogen-fixing symbiotic bacterium *Bradyrhizobium japonicum* [73-78]. The hairy root system was also used for analyzing roles of a MYB transcription factor in isoflavonoid biosynthesis [79].

Transgene-induced RNA silencing has also been induced in leaf tissues for the β-glucuronidase gene [80] or the senescence-associated receptor-like kinase gene [81] and in calli for the amino aldehyde dehydrogenase gene to induce the biosynthesis of 2-acetyl-1-pyrroline [82].

4.2. Disease resistance acquired by transgene-induced RNA silencing

Another focus of modifying soybean plants through RNA silencing is resistance against diseases, particularly to those caused by viruses. Resistance to viruses was achieved by transforming plants with genes or segments of genes derived from viruses and was referred to as pathogen-derived resistance [23, 24, 83, 84]. The resistance did not need protein translated from the transgene [85-87], which led to the understanding that RNA is the factor that conferred resistance to the plants and that the enhanced resistance is acquired via a mechanism analogous to that involved in co-suppression. Using this strategy, soybean plants resistant to *Soybean mosaic virus* (SMV) [88-90], or *Soybean dwarf virus* [91, 92] have been produced.

In addition to resistance against a virus, transgenic soybean plants resistant to cyst nematode (*Heterodera glycines*) have also been produced using RNA silencing [93], in which an inverted repeat of the major sperm protein gene from cyst nematode was transcribed from the transgene. RNA silencing was elicited in cyst nematode after nematode ingestion of dsRNA molecules produced in the soybean plants; as a consequence, reproductive capabilities of the cyst nematode were suppressed. The effects of RNA silencing on controlling *H. glycines* [94] or root-knot nematode (*Meloidogyne incognita*) [95] infection have been assayed in soybean roots using *A. rhizogenes*-mediated transformation. On the other hand, this root transformation method has also been used for analyzing a role of host genes in resistance against diseases caused by *Phytophthora sojae* [96, 97], *Fusarium solani* [98] or cyst nematode [99].

4.3. Gene functional analysis by VIGS

An advantage of VIGS is its ease for making a gene construct and introducing nucleic acids to cells. In addition, the effect of silencing can be monitored within a short time after inoculating plants with the virus. Because of these features, VIGS is suitable for gene function analysis [51, 100, 101] and has been used for gene identification via downregulating a candidate gene(s) responsible for a specific phenomenon in soybean. VIGS was used to demonstrate that genes present in the genetically identified loci actually encode the genes responsible for the phenotype: VIGS of the putative *flavonoid 3'-hydroxylase* (F3'H) gene resulted in a decrease in the content of quercetin relative to kampferol, which indicated that the putative gene actually encodes the F3'H protein [56]; VIGS of the *GmTFL1b* gene, a soybean orthologue of *Arabidopsis TERMINAL FLOWER1* (TFL1) and a candidate gene for the genetically identified locus *Dt1*, induced an early transition from vegetative to reproductive phases, which indicated the identity between *Dt1* and *GmTFL1b* [102]. VIGS has also been used to identify genes involved in resistance of soybean plants against pathogens such as SMV, BPMV, *Pseudomonas syringae* or *Phakopsora pachyrhizi* [103-107].

4.4. Naturally occurring RNA silencing

In addition to artificially induced RNA silencing, naturally occurring RNA silencing has also been known in soybean. Naturally occurring RNA silencing, involving mRNA degradation induced as a consequence of certain genetic changes, has been detected based on phenotypic changes. Most commercial varieties of soybean produce yellow seeds due to loss of pigmentation in seed coats, and this phenotype has been shown to be due to PTGS of the *CHS* genes [108, 109]. In cultivated soybean, there are varieties producing seeds with yellow seed coats and those producing seeds with brown or black seed coats in which anthocyanin and proanthocyanidin accumulate. In contrast, wild soybeans (*Glycine soja*), an ancestor of the cultivated soybean, have exclusively produced seeds with pigmented seed coats in thousands of accessions from natural populations in East Asia that we have screened (unpublished data). Thus, the nonpigmented seed coat phenotype was probably generated after domestication of soybean, and humans have maintained the plant lines that have *CHS* RNA silencing. The genetic change that induced *CHS* RNA silencing has been attributed to a structural change in the *CHS* gene cluster, which allows production of inverted repeat *CHS* RNA [110].

The occurrence of RNA silencing that leads to changes in pigmentation of plant tissues has also been reported for the *CHS* genes in maize [111] and petunia [112]. In petunia, a variety 'Red Star' produces flowers having a star-type red and white bicolor pattern, which resembles the flower-color patterns observed in transgenic petunias with co-suppression of the *CHS* genes [113], and in fact, the phenotype was demonstrated to be due to RNA silencing of the *CHS* genes in the white sectors [112]. Breeding of petunia was launched in the 1830s by crossing among wild species. The generation of the star-type petunia plants as a consequence of hybridizations between plant lines suggests that RNA silencing ability can be conferred via shuffling of genomes that are slightly different from each other. These phenomena also resemble the RNA silencing in a seed storage protein gene in rice, which is associated with a structural change in the gene region induced by mutagenesis [114], a case of RNA silencing in nontransgenic plants.

5. Diagnosis of an RNA silencing-induced phenotype using viral infection

In the course of the analysis of *CHS* RNA silencing, the function of a virus-encoded protein called suppressor protein of RNA silencing was used to visually demonstrate the occurrence of RNA silencing [108, 111, 112, 115]. These suppressor proteins affect viral accumulation in plants. The ability of the suppressor protein to allow viral accumulation is due to its inhibition of RNA silencing by preventing the incorporation of siRNAs into RISCs or by interfering with RISCs [116]. Because of these features, RNA silencing can be suppressed in plants infected with a virus that carries the suppressor protein. When a soybean plant that has a yellow seed coat is infected with CMV, the seed coat restores pigmentation [108]. This phenomenon is due to the activity of gene silencing suppressor protein called 2b encoded by the CMV. This example typically indicates that, using the function of viral suppressor protein we can "diagnose" whether an observed phenotypic change in a plant is caused by RNA silencing. A similar phenomenon has also been detected in maize [111] and petunia [112] lines, both of which have phenotypic changes through naturally occurring RNA silencing of an endogenous *CHS* gene, or a transgenic petunia line that has *CHS* co-suppression [115].

6. What do phenotypic changes induced by RNA silencing in soybean indicate?

Soybean is thought to be derived from an ancestral plant(s) with a tetraploid genome, and as a consequence, large portions of the soybean genome are duplicated [7], with nearly 75% of the genes present in multiple copies [117]. In addition, genes in the soybean genome are sometimes duplicated in tandem [118-121]. Our recent studies have indeed shown functional redundancy of duplicated genes in soybean [122, 123]. Such gene duplication can be an obstacle to producing mutants by conventional methods of mutagenesis. In this regard, the gene silencing technique is particularly useful because it allows silencing of multiple cognate genes having nucleotide sequence identity.

Changes in phenotypes as a consequence of inducing RNA silencing have been successful for many genes in soybean as mentioned above. Considering that many genes are duplicated in soybean genome, this fact indicates either that RNA silencing worked on all duplicated genes that have the same function or that the genes were subfunctionalized after duplication, so that RNA silencing of even a single gene of the duplicated genes resulted in the phenotypic changes.

It is of interest to understand whether duplicated genes have identical or diversified functions, which may depend on the time after duplication event and/or the selection pressure on the genes. To analyze the functions of each copy of the duplicated genes, we need to silence a specific copy of the duplicated genes. If the duplicated genes are expressed in different tissues, RNA silencing of both genes can lead to understanding the function of each gene. PTGS by transcribing inverted repeat with a constitutive promoter or VIGS will be suitable for this analysis. An example of such an approach is the VIGS of duplicated *TFL1* orthologues, which are expressed in different tissues. A specific role of one of the *TFL1* orthologues has been identified by VIGS as mentioned earlier [102].

7. Methods to induce selective RNA silencing of duplicated genes

When duplicated genes are subfunctionalized with only limited nucleotide changes and are expressed in overlapping tissues, specific silencing of each gene will be necessary for understanding their function(s). Silencing a specific copy of duplicated genes can be achieved by targeting a gene portion whose nucleotide sequence is differentiated between the duplicated genes. A condition that allows this type of silencing involves a lack of silencing of the other copy of duplicated genes even when they have the same sequence in the other portions.

In plants, miRNAs or siRNAs promote production of secondary siRNAs from the 5' up-stream region and/or the 3' downstream region of the initially targeted region via produc-tion of dsRNA by RdRP. These secondary siRNAs can lead to silencing of a secondary target that is not directly targeted by the primary silencing trigger [124]. Studies so far have indi-cated that such a spread of RNA silencing, called transitive RNA silencing, does not occur with the majority of endogenous genes, although it can happen to a transgene [45; and refer-ences therein]. Assuming the lack of transitive RNA silencing, it is possible to induce silenc-ing of a specific copy of a duplicated gene. Targeting a region specific for each copy, e.g., the 3' untranslated region (UTR), can induce silencing of the gene copy only, whereas targeting a region conserved in duplicated gene copies can induce silencing of the multiple gene copies simultaneously (Figure 3). Such selective RNA silencing was successful in a gene family of rice [125] and this strategy may work for analyzing functional diversification of duplicated genes in any plant species.

An alternative approach to suppress gene expression in plants is the use of artificial miR-NAs (Figure 4) (amiRNAs; also called synthetic miRNAs) [38, 126]. This approach involves modification of plant miRNA sequence to target specific transcripts, originally not under miRNA control, and downregulation of gene expression via specific cleavage of the target RNA. Melito *et al.* have used amiRNA to downregulate the leucine-rich repeat transmem-brane receptor-kinase gene in soybean [99]. miRNA has been extensively studied in soybean [127-130], information of which may be useful for designing amiRNAs. Because of its specif-icity, this method will be useful for silencing a limited copy of duplicated genes in soybean.

Induction of TGS by targeting dsRNA to a gene promoter can also be the method of choice. Gene silencing through transcriptional repression can be induced by dsRNA targeted to a gene promoter (Figure 4). However, until recently, no plant has been produced that harbors an endogenous gene that remains silenced in the absence of promoter-targeting dsRNA. We have reported for the first time that TGS can be induced by targeting dsRNA to the endoge-nous gene promoters in petunia and tomato plants, using a *Cucumber mosaic virus* (CMV)-based vector and that the induced gene silencing is heritable. Efficient silencing depended on the function of the 2b protein encoded in the vector, which facilitates epigenetic modifi-cations through the transport of siRNA to the nucleus [131, 132]. The progeny plants do not have any transgene because the virus is eliminated during meiosis. Therefore, plants that are produced by this system have altered traits but do not carry a transgene, thus constituting a novel class of modified plants [131, 132]. We have also developed *in planta* assay systems to detect inhibition of cytosine methylation using plants that contain a transgene transcription-ally silenced by an epigenetic mechanism [133]. Using these systems, we found that genis-tein, a major isoflavonoid compound rich in soybean seeds, inhibits cytosine methylation and restores the transcription of epigenetically silenced genes [133]. Whether developing soybean seeds are resistant (or susceptible) to epigenetic modifications is an interesting issue in terms of both developmental control of gene expression and intentionally inducing TGS through epigenetic changes.

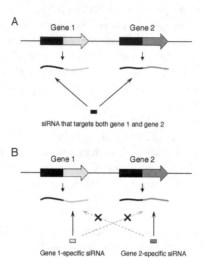

Figure 3. Selective RNA silencing of duplicated genes. The gene 1 and gene 2 are produced as a consequence of gene duplication. They share a highly conserved nucleotide sequence in the 5′ region, while they have a different sequence in the 3′ region. When siRNAs corresponding to the conserved region are produced, they can induce RNA degradation of the transcripts from both genes (A). On the other hand, siRNAs corresponding to the 3′ region can induce gene 1-specific or gene 2-specific RNA degradation (B). A combination of these different approaches enables functional analysis of duplicated genes.

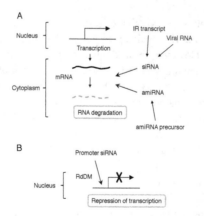

Figure 4. Various pathways of RNA silencing that can be intentionally induced to suppress gene expression in plants. Transcripts from transgenes that have an IR sequence of a plant gene segment or viral genomic RNA that carries the segment can form dsRNA. These dsRNAs are subsequently processed into siRNAs in the cytoplasm. Similarly, amiRNA precursors transcribed from the amiRNA gene are processed into amiRNAs. These small RNAs can cause degradation of target gene transcripts, a process termed PTGS (A). When siRNAs corresponding to a gene promoter are produced, they can induce RdDM in the nucleus, thereby TGS of the target gene can be induced (B). Abbreviations: amiRNA, artificial microRNA; PTGS, posttranscriptional gene silencing; RdDM, RNA-directed DNA methylation; TGS, transcriptional gene silencing.

8. Differentiation of duplicated genes and induction of RNA silencing

How much sequence difference will be necessary to induce selective RNA silencing? A factor that affects induction of RNA silencing is the extent of sequence identity between the dsRNA that triggers RNA silencing and its target gene. IR-PTGS could be induced by IR-transcripts that can form 98-nt or longer dsRNAs [39]. In VIGS, the lower size limit of the inserted fragments required for inducing PTGS is 23-nt, a size almost corresponding to that of siRNAs [134], and that for inducing TGS is 81-91 nt [135]. Silencing a gene probably requires sequence identity longer than the size of siRNAs between dsRNA and its target, although the efficiency of silencing may depend on the system of silencing induction.

We previously induced *CHS* VIGS in soybean [56]. In soybean seed coats, the *CHS7/CHS8* genes, which share 98% nucleotide sequence identity in the coding region, are predominantly expressed among the eight members of the *CHS* gene family [136, 137]. We have induced the silencing using a virus vector that carried a 244-nt fragment of the *CHS7* gene [56]. The *CHS* mRNA levels in the seed coats and leaf tissues of plants infected with the virus were reduced to 12.4% and 47.0% of the control plants, respectively. One plausible explanation for the differential effects of VIGS on these tissues may be that the limited sequence homology (79%-80%) between the *CHS7* and the *CHS1-CHS3* genes, the transcripts of which make up approximately 40% of the total *CHS* transcript content of leaf tissues [137], results in the degradation of the *CHS1-CHS3* transcripts at a lower efficiency than the degradation of *CHS7/ CHS8* transcripts. Consistent with these results, naturally occurring *CHS* RNA silencing, in which *CHS7/CHS8* genes are silenced in seed coat tissues, is thought to be induced by inverted repeat transcripts of a *CHS3* gene segment [110]. In terms of the practical use of transgene-induced RNA silencing, these results suggest that a portion of genes whose sequence identity between duplicated genes is lower than 79%-80% should be chosen as a target for inducing selective RNA silencing.

The naturally occurring RNA silencing of the *CHS* genes in soybean may indicate relationships between diversification of duplicated genes and RNA silencing. Gene duplication can be a cause of RNA silencing because it may sometimes result in the production of dsRNA, which triggers RNA silencing through read-through transcription [114, 115]. In the *CHS* silencing in soybean, the extent of mRNA decrease differs between different copies of the gene family. These observations may indicate that plants use subfunctionalization of duplicated genes as a means to avoid the occurrence of simultaneous silencing of duplicated genes, which may have a deleterious effect on the organism.

9. Conclusion and perspectives

RNA silencing has been used as a powerful tool to engineer novel traits or analyze gene function in soybean. Soybean plants that have engineered a metabolic pathway or acquired resistance to diseases have been produced by transgene-induced gene silencing. VIGS has been used as a tool to analyze gene function in soybean. In addition to RNA silencing, site-directed mutagenesis using zinc-finger nucleases has been applied to mutagenizing dupli-cated genes in soybean [138]. Such reverse genetic approaches may be supplemented by forward genetic approaches such as high linear energy transfer radiation-based mutagene-sis, e.g., irradiation of ion beam [139] and fast neutron [140]. Similarly, gene tagging systems using maize Ds transposon [141] and rice *mPing* transposon [142] have also been developed in soybean. Aside from using RNA silencing as a tool to engineer novel traits, analysis of mutants in combination with reverse genetic approaches may facilitate the identification of causative gene(s) of the mutation. An interesting feature of RNA silencing is its inducible nature, which allows downregulation of a gene in a tissue-specific manner. This strategy is particularly advantageous for analyzing the function of genes whose mutation or ubiquitous downregulation is lethal. Another feature of RNA silencing is that it allows analysis of bio-logical phenomena that involve the effect of a difference in the mRNA level of the gene. The dependence of pigmentation in soybean pubescence on the mRNA level of the *F3'H* gene has actually been shown by utilizing VIGS [143]. In this regard, selective RNA silencing of duplicated genes may reveal the presence of additive effects of the expression levels of du-plicated genes in soybean.

Abbreviations

AGO, Argonaute; ALSV, *Apple latent spherical virus*; amiRNA, artificial miRNA; BPMV, *Bean pod mottle virus*; CaMV, *Cauliflower mosaic virus*; CHS, chalcone synthase; CMV, *Cucumber mosaic virus*; dsRNA, double-stranded RNA; EST, expressed sequence tag; F3'H, flavonoid 3'-hydroxylase; IR, inverted repeat; miRNA, micro RNA; MYA, million years ago; PTGS, post-transcriptional gene silencing; RdRP, RNA-dependent RNA polymerase; RISC, RNA-induced silencing complex; RNAi, RNA interference; siRNA, short interfering RNA; SMV, *Soybean mosaic virus*; TFL1, TERMINAL FLOWER1; TGS, transcriptional gene silencing; UTR, untranslated region; VIGS, virus-induced gene silencing

Acknowledgements

Our work is supported by Grants-in-Aid for Scientific Research from the Ministry of Educa-tion, Culture, Sports, Science and Technology of Japan.

Author details

Megumi Kasai, Mayumi Tsuchiya and Akira Kanazawa*

*Address all correspondence to: kanazawa@res.agr.hokudai.ac.jp

Research Faculty of Agriculture, Hokkaido University, Japan

References

[1] Lackey, J. (1980). Chromosome-numbers in the Phaseoleae (Fabaceae, Faboideae) and their relation to taxonomy. *Am J Bot*, 67, 595-602.

[2] Hymowitz, T. (2004). Speciation and cytogenetics. In: Boerma HR, Specht JE. (eds) Soybeans: improvement, production, and uses, Ed 3, Agronomy Monograph No. 16. American Society of Agronomy, Inc., Crop Science Society of America, Inc., Soil Science Society of America, Inc. Madison, Wisconsin, USA; 2004. p97-136.

[3] Shoemaker, R. C., Schlueter, J., & Doyle, J. J. (2006). Paleopolyploidy and gene duplication in soybean and other legumes. *Curr Opin Plant Biol*, 9, 104-9.

[4] Moore, R. C., & Purugganan, M. D. (2005). The evolutionary dynamics of plant duplicate genes. *Curr Opin Plant Biol*, 8, 122-8.

[5] Lohnes, D., Specht, J., & Cregan, P. (1997). Evidence for homoeologous linkage groups in the soybean. *Crop Sci*, 37, 254-7.

[6] Zhu, T., Schupp, J. M., Oliphant, A., & Keim, P. (1994). Hypomethylated sequences: characterization of the duplicate soybean genome. *Mol Gen Genet*, 244, 638-45.

[7] Shoemaker, R. C., Polzin, K., Labate, J., Specht, J., Brummer, E. C., Olson, T., et al. (1996). Genome duplication in soybean (*Glycine* subgenus *soja*). *Genetics*, 144, 329-38.

[8] Lee, J., Bush, A., Specht, J., & Shoemaker, R. (1999). Mapping of duplicate genes in soybean. *Genome*, 42, 829-36.

[9] Schlueter, J. A., Dixon, P., Granger, C., Grant, D., Clark, L., Doyle, J. J., et al. (2004). Mining EST databases to resolve evolutionary events in major crop species. *Genome*, 47, 868-76.

[10] Lynch, M., & Conery, J. S. (2000). The evolutionary fate and consequences of duplicate genes. *Science*, 290, 1151-5.

[11] Force, A., Lynch, M., Pickett, F. B., Amores, A., Yan, Y. L., & Postlethwait, J. (1999). Preservation of duplicate genes by complementary, degenerative mutations. *Genetics*, 151, 1531-45.

[12] Hughes, A. L. (1994). The evolution of functionally novel proteins after gene duplication. *Proc Biol Sci*, 256, 119-24.

[13] Lynch, M., & Force, A. (2000). The probability of duplicate gene preservation by subfunctionalization. *Genetics*, 154, 459-73.

[14] Rodin, S. N., & Riggs, A. D. (2003). Epigenetic silencing may aid evolution by gene duplication. *J Mol Evol*, 56, 718-29.

[15] Freeling, M., & Thomas, B. C. (2006). Gene-balanced duplications, like tetraploidy, provide predictable drive to increase morphological complexity. *Genome Res*, 16, 805-14.

[16] Birchler, J. A., & Veitia, R. A. (2007). The gene balance hypothesis: from classical genetics to modern genomics. *Plant Cell*, 19, 395-402.

[17] Schlueter, J. A., Scheffler, B. E., Schlueter, S. D., & Shoemaker, R. C. (2006). Sequence conservation of homeologous bacterial artificial chromosomes and transcription of homeologous genes in soybean (*Glycine max* L. Merr.). *Genetics*, 174, 1017-28.

[18] Schlueter, J., Vaslenko-Sanders, I., Deshpande, S., Yi, J., Siegfried, M., Roe, B., et al. (2007). The FAD2 gene family of soybean: insights into the structural and functional divergence of a paleopolyploid genome. *Crop Sci*, 47, S14-S26.

[19] Brodersen, P., & Voinnet, O. (2006). The diversity of RNA silencing pathways in plants. *Trends Genet*, 22, 268-80.

[20] Vaucheret, H. (2006). Post-transcriptional small RNA pathways in plants: mechanisms and regulations. *Genes Dev*, 20, 759-71.

[21] Napoli, C., Lemieux, C., & Jorgensen, R. (1990). Introduction of a chimeric chalcone synthase gene into petunia results in reversible co-suppression of homologous genes *in trans*. *Plant Cell*, 2, 279-89.

[22] van der Krol, A. R., Mur, L. A., Beld, M., Mol, J. N., & Stuitje, A. R. (1990). Flavonoid genes in petunia: addition of a limited number of gene copies may lead to a suppression of gene expression. *Plant Cell*, 2, 291-9.

[23] Wilson, T. M. (1993). Strategies to protect crop plants against viruses: pathogen-derived resistance blossoms. *Proc Natl Acad Sci U S A*, 90, 3134-41.

[24] Baulcombe, D. C. (1996). Mechanisms of pathogen-derived resistance to viruses in transgenic plants. *Plant Cell*, 8, 1833-44.

[25] Metzlaff, M., O'Dell, M., Cluster, P. D., & Flavell, R. B. (1997). RNA-mediated RNA degradation and chalcone synthase A silencing in petunia. *Cell*, 88, 845-54.

[26] Waterhouse, P., Graham, H., & Wang, M. (1998). Virus resistance and gene silencing in plants can be induced by simultaneous expression of sense and antisense RNA. *Proc Natl Acad Sci U S A*, 95, 13959-64.

[27] Fire, A., Xu, S., Montgomery, M. K., Kostas, S. A., Driver, S. E., & Mello, C. C. (1998). Potent and specific genetic interference by double-stranded RNA in *Caenorhabditis elegans*. *Nature*, 391, 806-11.

[28] Matzke, M., Matzke, A. J., & Kooter, J. M. (2001). RNA: guiding gene silencing. *Science*, 293, 1080-3.

[29] Baulcombe, D. (2004). RNA silencing in plants. *Nature*, 431, 356-63.

[30] Matzke, M., Kanno, T., Daxinger, L., Huettel, B., & Matzke, A. J. (2009). RNA-mediated chromatin-based silencing in plants. *Curr Opin Cell Biol*, 21, 367-76.

[31] Dunoyer, P., Schott, G., Himber, C., Meyer, D., Takeda, A., Carrington, J. C., et al. (2010). Small RNA duplexes function as mobile silencing signals between plant cells. *Science*, 328, 912-6.

[32] Molnár, A., Melnyk, C. W., Bassett, A., Hardcastle, T. J., Dunn, R., & Baulcombe, D. C. (2010). Small silencing RNAs in plants are mobile and direct epigenetic modification in recipient cells. *Science*, 328, 872-5.

[33] Mallory, A. C., & Vaucheret, H. (2006). Functions of microRNAs and related small RNAs in plants. *Nat Genet*, 38(S31-6).

[34] Voinnet, O. (2002). RNA silencing: small RNAs as ubiquitous regulators of gene expression. *Curr Opin Plant Biol*, 5, 444-51.

[35] Matzke, M., Aufsatz, W., Kanno, T., Daxinger, L., Papp, I., Mette, M. F., et al. (2004). Genetic analysis of RNA-mediated transcriptional gene silencing. *Biochim Biophys Acta*, 1677, 129-41.

[36] Mansoor, S., Amin, I., Hussain, M., Zafar, Y., & Briddon, R. W. (2006). Engineering novel traits in plants through RNA interference. *Trends Plant Sci*, 11, 559-65.

[37] Kanazawa, A. (2008). RNA silencing manifested as visibly altered phenotypes in plants. *Plant Biotechnol*, 25, 423-35.

[38] Frizzi, A., & Huang, S. (2010). Tapping RNA silencing pathways for plant biotechnology. *Plant Biotechnol J*, 8, 655-77.

[39] Wesley, S. V., Helliwell, C. A., Smith, N. A., Wang, M. B., Rouse, D. T., Liu, Q., et al. (2001). Construct design for efficient, effective and high-throughput gene silencing in plants. *Plant J*, 27, 581-90.

[40] Helliwell, C. A., & Waterhouse, P. M. (2005). Constructs and methods for hairpin RNA-mediated gene silencing in plants. *Methods Enzymol*, 392, 24-35.

[41] Wagner, N., Mroczka, A., Roberts, P. D., Schreckengost, W., & Voelker, T. (2011). RNAi trigger fragment truncation attenuates soybean *FAD2-1* transcript suppression and yields intermediate oil phenotypes. *Plant Biotechnol J*, 9, 723-8.

[42] Ruiz, M. T., Voinnet, O., & Baulcombe, D. C. (1998). Initiation and maintenance of virus-induced gene silencing. *Plant Cell*, 10, 937-46.

[43] Metzlaff, M., O'Dell, M., Hellens, R., & Flavell, R. B. (2000). Developmentally and transgene regulated nuclear processing of primary transcripts of *chalcone synthase A* in petunia. *Plant J*, 23, 63-72.

[44] Hoffer, P., Ivashuta, S., Pontes, O., Vitins, A., Pikaard, C., Mroczka, A., et al. (2011). Posttranscriptional gene silencing in nuclei. *Proc Natl Acad Sci U S A*, 108, 409-14.

[45] Vermeersch, L., De Winne, N., & Depicker, A. (2010). Introns reduce transitivity proportionally to their length, suggesting that silencing spreads along the pre-mRNA. *Plant J*, 64, 392-401.

[46] Christie, M., Croft, L. J., & Carroll, B. J. (2011). Intron splicing suppresses RNA silencing in *Arabidopsis*. *Plant J*, 68, 159-67.

[47] Que, Q., Wang, H. Y., English, J. J., & Jorgensen, R. A. (1997). The frequency and degree of cosuppression by sense chalcone synthase transgenes are dependent on transgene promoter strength and are reduced by premature nonsense codons in the transgene coding sequence. *Plant Cell*, 9, 1357-68.

[48] Yamada, T., Takagi, K., & Ishimoto, M. (2012). Recent advances in soybean transformation and their application to molecular breeding and genomic analysis. *Breeding Sci*, 61, 480-94.

[49] Covey, S., AlKaff, N., Langara, A., & Turner, D. (1997). Plants combat infection by gene silencing. *Nature*, 385, 781-2.

[50] Al-Kaff, N. S., Covey, S. N., Kreike, M. M., Page, A. M., Pinder, R., & Dale, P. J. (1998). Transcriptional and posttranscriptional plant gene silencing in response to a pathogen. *Science*, 279, 2113-5.

[51] Lu, R., Martin-Hernandez, A. M., Peart, J. R., Malcuit, I., & Baulcombe, D. C. (2003). Virus-induced gene silencing in plants. *Methods*, 30, 296-303.

[52] Molnár, A., Csorba, T., Lakatos, L., Várallyay, E., Lacomme, C., & Burgyán, J. (2005). Plant virus-derived small interfering RNAs originate predominantly from highly structured single-stranded viral RNAs. *J Virol*, 79, 7812-8.

[53] Kumagai, M. H., Donson, J., della -Cioppa, G., Harvey, D., Hanley, K., & Grill, L. K. (1995). Cytoplasmic inhibition of carotenoid biosynthesis with virus-derived RNA. *Proc Natl Acad Sci U S A*, 92, 1679-83.

[54] Purkayastha, A., & Dasgupta, I. (2009). Virus-induced gene silencing: a versatile tool for discovery of gene functions in plants. *Plant Physiol Biochem*, 47, 967-76.

[55] Zhang, C., & Ghabrial, S. A. (2006). Development of *Bean pod mottle virus*-based vectors for stable protein expression and sequence-specific virus-induced gene silencing in soybean. *Virology*, 344, 401-11.

[56] Nagamatsu, A., Masuta, C., Senda, M., Matsuura, H., Kasai, A., Hong, J. S., et al. (2007). Functional analysis of soybean genes involved in flavonoid biosynthesis by virus-induced gene silencing. *Plant Biotechnol J*, 5, 778-90.

[57] Yamagishi, N., & Yoshikawa, N. (2009). Virus-induced gene silencing in soybean seeds and the emergence stage of soybean plants with *Apple latent spherical virus* vectors. *Plant Mol Biol*, 71, 15-24.

[58] Kasai, M., & Kanazawa, A. (2012). RNA silencing as a tool to uncover gene function and engineer novel traits in soybean. *Breeding Sci*, 61, 468-79.

[59] Kinney, A. J., Jung, R., & Herman, E. M. (2001). Cosuppression of the α subunits of β-conglycinin in transgenic soybean seeds induces the formation of endoplasmic reticulum-derived protein bodies. *Plant Cell*, 13, 1165-78.

[60] Schmidt, M. A., Barbazuk, W. B., Sandford, M., May, G., Song, Z., Zhou, W., et al. (2011). Silencing of soybean seed storage proteins results in a rebalanced protein composition preserving seed protein content without major collateral changes in the metabolome and transcriptome. *Plant Physiol*, 156, 330-45.

[61] Nunes, A. C., Vianna, G. R., Cuneo, F., Amaya-Farfán, J., de Capdeville, G., Rech, E. L., et al. (2006). RNAi-mediated silencing of the *myo*-inositol-1-phosphate synthase gene (*GmMIPS1*) in transgenic soybean inhibited seed development and reduced phytate content. *Planta*, 224, 125-32.

[62] Shi, J., Wang, H., Schellin, K., Li, B., Faller, M., Stoop, J. M., et al. (2007). Embryo-specific silencing of a transporter reduces phytic acid content of maize and soybean seeds. *Nat Biotechnol*, 25, 930-7.

[63] Takagi, K., Nishizawa, K., Hirose, A., Kita, A., & Ishimoto, M. (2011). Manipulation of saponin biosynthesis by RNA interference-mediated silencing of β-amyrin synthase gene expression in soybean. *Plant Cell Rep*, 30, 1835-46.

[64] Herman, E. M., Helm, R. M., Jung, R., & Kinney, A. J. (2003). Genetic modification removes an immunodominant allergen from soybean. *Plant Physiol*, 132, 36-43.

[65] Yu, O., Shi, J., Hession, A. O., Maxwell, C. A., McGonigle, B., & Odell, J. T. (2003). Metabolic engineering to increase isoflavone biosynthesis in soybean seed. *Phytochemistry*, 63, 753-63.

[66] Kinney, A. (1996). Development of genetically engineered soybean oils for food applications. *J Food Lipids*, 3, 273-92.

[67] Chen, R., Matsui, K., Ogawa, M., Oe, M., Ochiai, M., Kawashima, H., et al. (2006). Expression of Δ6, Δ5 desaturase and GLELO elongase genes from *Mortierella alpina* for production of arachidonic acid in soybean [*Glycine max* (L.) Merrill] seeds. *Plant Sci*, 170, 399-406.

[68] Flores, T., Karpova, O., Su, X., Zeng, P., Bilyeu, K., Sleper, D. A., et al. (2008). Silencing of *GmFAD3* gene by siRNA leads to low alpha-linolenic acids (18:3) of fad3-mutant phenotype in soybean [*Glycine max* (Merr.)]. *Transgenic Res*, 17, 839-50.

[69] Schmidt, M. A., & Herman, E. M. (2008). Suppression of soybean oleosin produces micro-oil bodies that aggregate into oil body/ER complexes. *Mol Plant*, 1, 910-24.

[70] Wang, G., & Xu, Y. (2008). Hypocotyl-based Agrobacterium-mediated transformation of soybean (*Glycine max*) and application for RNA interference. *Plant Cell Rep*, 27, 1177-84.

[71] Lee, J., Welti, R., Schapaugh, W. T., & Trick, H. N. (2011). Phospholipid and triacylglycerol profiles modified by PLD suppression in soybean seed. *Plant Biotechnol J*, 9, 359-72.

[72] Wagner, N., Mroczka, A., Roberts, P. D., Schreckengost, W., & Voelker, T. (2011). RNAi trigger fragment truncation attenuates soybean *FAD2-1* transcript suppression and yields intermediate oil phenotypes. *Plant Biotechnol J*, 9, 723-8.

[73] Lee, M. Y., Shin, K. H., Kim, Y. K., Suh, J. Y., Gu, Y. Y., Kim, M. R., et al. (2005). Induction of thioredoxin is required for nodule development to reduce reactive oxygen species levels in soybean roots. *Plant Physiol*, 139, 1881-9.

[74] Subramanian, S., Stacey, G., & Yu, O. (2006). Endogenous isoflavones are essential for the establishment of symbiosis between soybean and *Bradyrhizobium japonicum*. *Plant J*, 48, 261-73.

[75] Hayashi, S., Gresshoff, P. M., & Kinkema, M. (2008). Molecular analysis of lipoxygenases associated with nodule development in soybean. *Mol Plant Microbe Interact*, 21, 843-53.

[76] Dalton, D. A., Boniface, C., Turner, Z., Lindahl, A., Kim, H. J., Jelinek, L., et al. (2009). Physiological roles of glutathione S-transferases in soybean root nodules. *Plant Physiol*, 150, 521-30.

[77] Govindarajulu, M., Kim, S. Y., Libault, M., Berg, R. H., Tanaka, K., Stacey, G., et al. (2009). GS52 ecto-apyrase plays a critical role during soybean nodulation. *Plant Physiol*, 149, 994-1004.

[78] Libault, M., Zhang, X. C., Govindarajulu, M., Qiu, J., Ong, Y. T., Brechenmacher, L., et al. (2010). A member of the highly conserved *FWL* (tomato *FW2.2-like*) gene family is essential for soybean nodule organogenesis. *Plant J*, 62, 852-64.

[79] Yi, J., Derynck, M. R., Li, X., Telmer, P., Marsolais, F., & Dhaubhadel, S. (2010). A single-repeat MYB transcription factor, GmMYB176, regulates *CHS8* gene expression and affects isoflavonoid biosynthesis in soybean. *Plant J*, 62, 1019-34.

[80] Reddy, M. S., Dinkins, R. D., & Collins, G. B. (2003). Gene silencing in transgenic soybean plants transformed via particle bombardment. *Plant Cell Rep*, 21, 676-83.

[81] Li, X. P., Gan, R., Li, P. L., Ma, Y. Y., Zhang, L. W., Zhang, R., et al. (2006). Identification and functional characterization of a leucine-rich repeat receptor-like kinase gene that is involved in regulation of soybean leaf senescence. *Plant Mol Biol*, 61, 829-44.

[82] Arikit, S., Yoshihashi, T., Wanchana, S., Uyen, T. T., Huong, N. T., Wongpornchai, S., et al. (2011). Deficiency in the amino aldehyde dehydrogenase encoded by *GmAMADH2*, the homologue of rice *Os2AP*, enhances 2-acetyl-1-pyrroline biosynthesis in soybeans (*Glycine max* L.). *Plant Biotechnol J*, 9, 75-87.

[83] Prins, M., & Goldbach, R. (1996). RNA-mediated virus resistance in transgenic plants. *Arch Virol*, 141, 2259-76.

[84] Goldbach, R., Bucher, E., & Prins, M. (2003). Resistance mechanisms to plant viruses: an overview. *Virus Res*, 92, 207-12.

[85] Smith, H. A., Swaney, S. L., Parks, T. D., Wernsman, E. A., & Dougherty, W. G. (1994). Transgenic plant virus resistance mediated by untranslatable sense RNAs: expression, regulation, and fate of nonessential RNAs. *Plant Cell*, 6, 1441-53.

[86] Mueller, E., Gilbert, J., Davenport, G., Brigneti, G., & Baulcombe, D. (1995). Homology-dependent resistance: transgenic virus resistance in plants related to homology-dependent gene silencing. *Plant J*, 7, 1001-13.

[87] Sijen, T., Wellink, J., Hiriart, J. B., & Van Kammen, A. (1996). RNA-mediated virus resistance: role of repeated transgenes and delineation of targeted regions. *Plant Cell*, 8, 2277-94.

[88] Wang, X., Eggenberger, A., Nutter, F., & Hill, J. (2001). Pathogen-derived transgenic resistance to soybean mosaic virus in soybean. *Mol Breed*, 8, 119-27.

[89] Furutani, N., Hidaka, S., Kosaka, Y., Shizukawa, Y., & Kanematsu, S. (2006). Coat protein gene-mediated resistance to soybean mosaic virus in transgenic soybean. *Breeding Sci*, 56, 119-24.

[90] Furutani, N., Yamagishi, N., Hidaka, S., Shizukawa, Y., Kanematsu, S., & Kosaka, Y. (2007). Soybean mosaic virus resistance in transgenic soybean caused by posttranscriptional gene silencing. *Breeding Sci*, 57, 123-8.

[91] Tougou, M., Furutani, N., Yamagishi, N., Shizukawa, Y., Takahata, Y., & Hidaka, S. (2006). Development of resistant transgenic soybeans with inverted repeat-coat protein genes of soybean dwarf virus. *Plant Cell Rep*, 25, 1213-8.

[92] Tougou, M., Yamagishi, N., Furutani, N., Shizukawa, Y., Takahata, Y., & Hidaka, S. (2007). Soybean dwarf virus-resistant transgenic soybeans with the sense coat protein gene. *Plant Cell Rep*, 26, 1967-75.

[93] Steeves, R., Todd, T., Essig, J., & Trick, H. (2006). Transgenic soybeans expressing siRNAs specific to a major sperm protein gene suppress *Heterodera glycines* reproduction. *Funct Plant Biol*, 33, 991-9.

[94] Li, J., Todd, T. C., Oakley, T. R., Lee, J., & Trick, H. N. (2010). Host-derived suppression of nematode reproductive and fitness genes decreases fecundity of *Heterodera glycines* Ichinohe. *Planta*, 232, 775-85.

[95] Ibrahim, H. M., Alkharouf, N. W., Meyer, S. L., Aly, M. A., Gamal El-Din, AlK., Hussein, E. H., et al. (2011). Post-transcriptional gene silencing of root-knot nematode in transformed soybean roots. *Exp Parasitol*, 127, 90-9.

[96] Subramanian, S., Graham, M. Y., Yu, O., & Graham, T. L. (2005). RNA interference of soybean isoflavone synthase genes leads to silencing in tissues distal to the transformation site and to enhanced susceptibility to *Phytophthora sojae*. *Plant Physiol*, 137, 1345-53.

[97] Graham, T. L., Graham, M. Y., Subramanian, S., & Yu, O. (2007). RNAi silencing of genes for elicitation or biosynthesis of 5-deoxyisoflavonoids suppresses race-specific resistance and hypersensitive cell death in *Phytophthora sojae* infected tissues. *Plant Physiol*, 144, 728-40.

[98] Lozovaya, V. V., Lygin, A. V., Zernova, O. V., Ulanov, A. V., Li, S., Hartman, G. L., et al. (2007). Modification of phenolic metabolism in soybean hairy roots through down regulation of chalcone synthase or isoflavone synthase. *Planta*, 225, 665-79.

[99] Melito, S., Heuberger, A. L., Cook, D., Diers, B. W., MacGuidwin, A. E., & Bent, A. F. (2010). A nematode demographics assay in transgenic roots reveals no significant impacts of the *Rhg1* locus LRR-Kinase on soybean cyst nematode resistance. *BMC Plant Biol*, 10, 104.

[100] Metzlaff, M. (2002). RNA-mediated RNA degradation in transgene- and virus-induced gene silencing. *Biol Chem*, 383, 1483-9.

[101] Burch-Smith, T. M., Anderson, J. C., Martin, G. B., & Dinesh-Kumar, S. P. (2004). Applications and advantages of virus-induced gene silencing for gene function studies in plants. *Plant J*, 39, 734-46.

[102] Liu, B., Watanabe, S., Uchiyama, T., Kong, F., Kanazawa, A., Xia, Z., et al. (2010). The soybean stem growth habit gene *Dt1* is an ortholog of Arabidopsis *TERMINAL FLOWER1*. *Plant Physiol*, 153, 198-210.

[103] Kachroo, A., Fu, D. Q., Havens, W., Navarre, D., Kachroo, P., & Ghabrial, S. A. (2008). An oleic acid-mediated pathway induces constitutive defense signaling and enhanced resistance to multiple pathogens in soybean. *Mol Plant Microbe Interact*, 21, 564-75.

[104] Fu, D. Q., Ghabrial, S., & Kachroo, A. (2009). *GmRAR1* and *GmSGT1* are required for basal, R gene-mediated and systemic acquired resistance in soybean. *Mol Plant Microbe Interact*, 22, 86-95.

[105] Meyer, J. D., Silva, D. C., Yang, C., Pedley, K. F., Zhang, C., van de Mortel, M., et al. (2009). Identification and analyses of candidate genes for *Rpp4*-mediated resistance to Asian soybean rust in soybean. *Plant Physiol*, 150, 295-307.

[106] Pandey, A. K., Yang, C., Zhang, C., Graham, M. A., Horstman, H. D., Lee, Y., et al. (2011). Functional analysis of the Asian soybean rust resistance pathway mediated by *Rpp2*. *Mol Plant Microbe Interact*, 24, 194-206.

[107] Singh, A. K., Fu, D. Q., El-Habbak, M., Navarre, D., Ghabrial, S., & Kachroo, A. (2011). Silencing genes encoding omega-3 fatty acid desaturase alters seed size and accumulation of *Bean pod mottle virus* in soybean. *Mol Plant Microbe Interact*, 24, 506-15.

[108] Senda, M., Masuta, C., Ohnishi, S., Goto, K., Kasai, A., Sano, T., et al. (2004). Patterning of virus-infected *Glycine max* seed coat is associated with suppression of endogenous silencing of chalcone synthase genes. *Plant Cell*, 16, 807-18.

[109] Senda, M., Kurauchi, T., Kasai, A., & Ohnishi, S. (2012). Suppressive mechanism of seed coat pigmentation in yellow soybean. *Breeding Sci*, 61, 523-30.

[110] Kasai, A., Kasai, K., Yumoto, S., & Senda, M. (2007). Structural features of GmIRCHS, candidate of the I gene inhibiting seed coat pigmentation in soybean: implications for inducing endogenous RNA silencing of chalcone synthase genes. *Plant Mol Biol*, 64, 467-79.

[111] Della Vedova, C. B., Lorbiecke, R., Kirsch, H., Schulte, M. B., Scheets, K., Borchert, L. M., et al. (2005). The dominant inhibitory chalcone synthase allele *C2-Idf* (*Inhibitor diffuse*) from *Zea mays* (L.) acts via an endogenous RNA silencing mechanism. *Genetics*, 170, 1989-2002.

[112] Koseki, M., Goto, K., Masuta, C., & Kanazawa, A. (2005). The star-type color pattern in *Petunia hybrida* 'Red Star' flowers is induced by sequence-specific degradation of chalcone synthase RNA. *Plant Cell Physiol*, 46, 1879-83.

[113] Jorgensen, R. A. (1995). Cosuppression, flower color patterns, and metastable gene expression States. *Science*, 268, 686-91.

[114] Kusaba, M., Miyahara, K., Iida, S., Fukuoka, H., Takano, T., Sassa, H., et al. (2003). *Low glutelin content1*: a dominant mutation that suppresses the *glutelin* multigene family via RNA silencing in rice. *Plant Cell*, 15, 1455-67.

[115] Kasai, M., Koseki, M., Goto, K., Masuta, C., Ishii, S., Hellens, R. P., et al. (2012). Coincident sequence-specific RNA degradation of linked transgenes in the plant genome. *Plant Mol Biol*, 78, 259-73.

[116] Silhavy, D., & Burgyán, J. (2004). Effects and side-effects of viral RNA silencing suppressors on short RNAs. *Trends Plant Sci*, 9, 76-83.

[117] Schmutz, J., Cannon, S. B., Schlueter, J., Ma, J., Mitros, T., Nelson, W., et al. (2010). Genome sequence of the palaeopolyploid soybean. *Nature*, 463, 178-83.

[118] Yoshino, M., Kanazawa, A., Tsutsumi, K., Nakamura, I., Takahashi, K., & Shimamoto, Y. (2002). Structural variation around the gene encoding the α subunit of soybean β-conglycinin and correlation with the expression of the α subunit. *Breeding Sci*, 52, 285-92.

[119] Matsumura, H., Watanabe, S., Harada, K., Senda, M., Akada, S., Kawasaki, S., et al. (2005). Molecular linkage mapping and phylogeny of the chalcone synthase multigene family in soybean. *Theor Appl Genet*, 110, 1203-9.

[120] Schlueter, J. A., Scheffler, B. E., Jackson, S., & Shoemaker, R. C. (2008). Fractionation of synteny in a genomic region containing tandemly duplicated genes across *Glycine max*, *Medicago truncatula*, and *Arabidopsis thaliana*. *J Hered*, 99, 390-5.

[121] Kong, F., Liu, B., Xia, Z., Sato, S., Kim, B. M., Watanabe, S., et al. (2010). Two coordinately regulated homologs of *FLOWERING LOCUS T* are involved in the control of photoperiodic flowering in soybean. *Plant Physiol*, 154, 1220-31.

[122] Liu, B., Kanazawa, A., Matsumura, H., Takahashi, R., Harada, K., & Abe, J. (2008). Genetic redundancy in soybean photoresponses associated with duplication of the phytochrome A gene. *Genetics, 180*, 995-1007.

[123] Kanazawa, A., Liu, B., Kong, F., Arase, S., & Abe, J. (2009). Adaptive evolution involving gene duplication and insertion of a novel *Ty1/copia*-like retrotransposon in soybean. *J Mol Evol, 69*, 164-75.

[124] Voinnet, O. (2008). Use, tolerance and avoidance of amplified RNA silencing by plants. *Trends Plant Sci, 13*, 317-28.

[125] Miki, D., Itoh, R., & Shimamoto, K. (2005). RNA silencing of single and multiple members in a gene family of rice. *Plant Physiol, 138*, 1903-13.

[126] Ossowski, S., Schwab, R., & Weigel, D. (2008). Gene silencing in plants using artificial microRNAs and other small RNAs. *Plant J, 53*, 674-90.

[127] Zhang, B., Pan, X., & Stellwag, E. J. (2008). Identification of soybean microRNAs and their targets. *Planta, 229*, 161-82.

[128] Chen, R., Hu, Z., & Zhang, H. (2009). Identification of microRNAs in wild soybean (*Glycine soja*). *J Integr Plant Biol, 51*, 1071-9.

[129] Song, Q. X., Liu, Y. F., Hu, X. Y., Zhang, W. K., Ma, B., Chen, S. Y., et al. (2011). Identification of miRNAs and their target genes in developing soybean seeds by deep sequencing. *BMC Plant Biol, 11*, 5.

[130] Turner, M., Yu, O., & Subramanian, S. (2012). Genome organization and characteristics of soybean microRNAs. *BMC Genomics, 13*, 169.

[131] Kanazawa, A., Inaba, J., Shimura, H., Otagaki, S., Tsukahara, S., Matsuzawa, A., et al. (2011). Virus-mediated efficient induction of epigenetic modifications of endogenous genes with phenotypic changes in plants. *Plant J, 65*, 156-68.

[132] Kanazawa, A., Inaba, J., Kasai, M., Shimura, H., & Masuta, C. R. N. (2011). RNA-mediated epigenetic modifications of an endogenous gene targeted by a viral vector: A potent gene silencing system to produce a plant that does not carry a transgene but has altered traits. *Plant Signal Behav, 6*, 1090-3.

[133] Arase, S., Kasai, M., & Kanazawa, A. (2012). *In planta* assays involving epigenetically silenced genes reveal inhibition of cytosine methylation by genistein. *Plant Methods, 8*, 10.

[134] Thomas, C. L., Jones, L., Baulcombe, D. C., & Maule, A. J. (2001). Size constraints for targeting post-transcriptional gene silencing and for RNA-directed methylation in *Nicotiana benthamiana* using a potato virus X vector. *Plant J, 25*, 417-25.

[135] Otagaki, S., Kawai, M., Masuta, C., & Kanazawa, A. (2011). Size and positional effects of promoter RNA segments on virus-induced RNA-directed DNA methylation and transcriptional gene silencing. *Epigenetics, 6*, 681-91.

[136] Kasai, A., Watarai, M., Yumoto, S., Akada, S., Ishikawa, R., Harada, T., et al. (2004). Influence of PTGS on chalcone synthase gene family in yellow soybean seed coat. *Breeding Sci*, 54, 355-60.

[137] Tuteja, J. H., Clough, S. J., Chan, W. C., & Vodkin, L. O. (2004). Tissue-specific gene silencing mediated by a naturally occurring chalcone synthase gene cluster in *Glycine max*. *Plant Cell*, 16, 819-35.

[138] Curtin, S. J., Zhang, F., Sander, J. D., Haun, W. J., Starker, C., Baltes, N. J., et al. (2011). Targeted mutagenesis of duplicated genes in soybean with zinc-finger nucleases. *Plant Physiol*, 156, 466-73.

[139] Arase, S., Hase, Y., Abe, J., Kasai, M., Yamada, T., Kitamura, K., et al. (2011). Optimization of ion-beam irradiation for mutagenesis in soybean: effects on plant growth and production of visibly altered mutants. *Plant Biotechnol*, 28, 323-9.

[140] Bolon, Y. T., Haun, W. J., Xu, W. W., Grant, D., Stacey, M. G., Nelson, R. T., et al. (2011). Phenotypic and genomic analyses of a fast neutron mutant population resource in soybean. *Plant Physiol*, 156, 240-53.

[141] Mathieu, M., Winters, E. K., Kong, F., Wan, J., Wang, S., Eckert, H., et al. (2009). Establishment of a soybean (*Glycine max* Merr. L) transposon-based mutagenesis repository. *Planta*, 229, 279-89.

[142] Hancock, C. N., Zhang, F., Floyd, K., Richardson, A. O., Lafayette, P., Tucker, D., et al. (2011). The rice miniature inverted repeat transposable element *mPing* is an effective insertional mutagen in soybean. *Plant Physiol*, 157, 552-62.

[143] Nagamatsu, A., Masuta, C., Matsuura, H., Kitamura, K., Abe, J., & Kanazawa, A. (2009). Down-regulation of flavonoid 3'-hydroxylase gene expression by virus-induced gene silencing in soybean reveals the presence of a threshold mRNA level associated with pigmentation in pubescence. *J Plant Physiol*, 166, 32-9.

Chapter 6

Proteomics and Its Use in Obtaining Superior Soybean Genotypes

Cristiane Fortes Gris and Alexana Baldoni

Additional information is available at the end of the chapter

1. Introduction

Soybean (*Glycine max* L. Merrill) is one of the most important and most cultivated crops in the world, with significant quantities of proteins being found in their yield composition, around 40% of their yield dry matter. This expressive quantity of proteins, and also a considerable percentage of oil, around 21% of their dry matter, has turned this grain into a product of great importance for the industrial sector, whether it be for food, cosmetics or, more recently, biofuels. Thus, soybean breeding programs directed toward these areas become ever more important, together with agronomic characteristics that allow greater productivity in sustainability with the environment in which they are produced.

The achievement of soybean genome sequencing [1], facilitated by identification of the genetic base, lead to advances in obtaining improved cultivars through knowledge of the complete sequence of expressed genes. Nevertheless, this information is not sufficient to identify which proteins are really being expressed in the cell at a given moment and under a certain condition since, through the phenomenon of splicing, different proteins may be produced by alteration of the command of a single gene. Thus, the complementary DNA (cDNA) and the messenger RNA (mRNA) have come to be the main focus of study for obtaining information regarding genetic expression or transcriptome. Nevertheless, due to post-translational regulation mechanisms, the quantity of expressed protein is not necessarily proportional to the quantity of its corresponding mRNA, which often raises questions regarding the role of this gene in cellular metabolism.

The reason for this is that control of gene expression occurs from mRNA transcription up to post-translational modifications like glycosylation and phosphorylation, among other processes, which alter protein activity (Figure 1).

In recent years, for the purpose of complementing the information obtained by means of genome sequencing and transcriptome, proteomics, one of the dimensions of the post-genome era [2], arises with a set of highly powerful techniques for separation and identification of proteins in biological samples, allowing better understanding of the networks of cellular operation and regulation upon representing the link between the genotype and the phenotype of an organism.

For the aforementioned reasons, proteomic analysis is now one of the most efficient means for functional study of the genes and genomes of complex organisms [3]. This has generated new data, as well as validated, complemented and even corrected information obtained through other approaches, thus contributing to better understanding of plant biology.

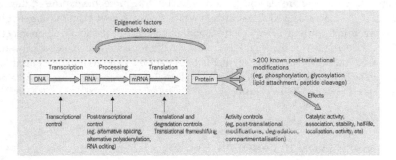

Figure 1. Pathways in which gene and protein expression may be regulated or modified in transcription or in post-translation [13].

Its study involves the entire set of proteins expressed by the genome of a cell, or only those that are expressed differentially under specific conditions. Also it is directed to the set of protein isoforms and post-translational modifications, to the interactions among them, as well as to the structural description of molecules and their complexes.

Bidimensional electrophoresis and mass spectrometry are the core technologies of proteomics, although new methodologies are being applied to plants for specific studies [4,5,6]. Among the most recent proteomic techniques are Difference Gel Electrophoresis (DIGE) and Multi-dimensional Protein Identification Tecnology (MudPIT), used in separation of proteins from a complex mixture. Other methods involved are Stable Isotopic Labeling using Amino Acids in Cell Culture (SILAC), Isotope Coded Affinity Tag (ICAT) and Isobaric Tag for Relative and Absolute Quantitation (iTRAQ) are based on labeling with isotopes for quantification of molecules by mass spectrometry.

In spite of the recent nature of research in this area, diverse studies with soybeans using proteomic tools are being performed throughout the world, showing this to be a promising area for selection of genotypes for genetic breeding programs [7,8]. Moreover, the study of plant responses to infections from pathogens has supplied significant data for understanding the signaling process that triggers the defense response in plants [9]. Additionally, there are

studies characterizing the proteome of plants in response to different stress conditions aris-
ing from both abiotic factors [10] and biotic factors [11]. These comparative studies of contrasting genotypes for a determined type of stress allow identification of the proteins that
respond to stress by means of changes in their levels of expression. Identifying these molecules and their respective functions, the work of breeding is directed and should have con-
tinuity only with those molecules that perform roles related to the characteristic of stress
tolerance. For that reason, it is essential to cross the proteomic data with information also
obtained by genomics, transcriptomics and metabolomics, so as to verify the correlation of
the candidate proteins with the desired characteristic.

In relation to products derived from genetically modified foods, proteomic techniques have
been applied to allow a broad approach and the analysis of many variables simultaneously
in a single sample. There are also other studies relating the proteome expressed during development of the plants, as well as research in which soybeans have been the target of investigations regarding nutritional, toxicological and allergenic aspects, above all on genetically
modified varieties [12].This makes for increased use of this technique in biosecurity studies.
In this context, the objective of this chapter is to present the main technologies used in proteomic studies in diverse areas of activity, as well as the main scientific results obtained in
the search for superior soybean genotypes.

2. Technologies used in proteomic studies.

Execution of a proteomic study involves the integration of many technologies which perme-
ate the fields of molecular biology, biochemistry, physiology, statistics and bioinformatics,
among other areas. The key steps in this type of study are separation of complex mixtures of
proteins and their identification.

Separation is performed through the use of electrophoresis a term created by Michaelis in
1909. The first electrophoresis of proteins (Figure 2) was performed in 1937. Alfenas (1998)
[14] explains that electrophoresis aims at separation of molecules in terms of their electrical
charges, their molecular weights and their conformations, in porous supports and appropriate buffers, under the influence of a continuous electrical field. Molecules with a preponder-
ance of negative charges migrate in the electrical field to the positive pole (anode), and
molecules with excess of positive charges migrate to the negative pole (cathode). The pre-
ponderant charge of a proteic molecule is in accordance with its amino acids.

Many of the technologies currently used in proteomics were developed much before the beginning of proteomics, as is the case of electrophoresis. Nevertheless, it was the advance in
protein sequencing technology by means of mass spectrometry that allowed its emergence
and development [15].

The study of proteomics may be performed by means of techniques like two-dimensional
electrophoresis in polyacrylamide gel (2D PAGE) followed by mass spectrometry (MS) (Figure 3), or furthermore, more recently, by the association of ionization and chromatographic

methods, among others, which increase detection sensitivity even more. Nevertheless, the point of departure has still been the exposure of a large number of proteins from a cell line or organism in two-dimensional polyacrylamide gels [16,17,18].

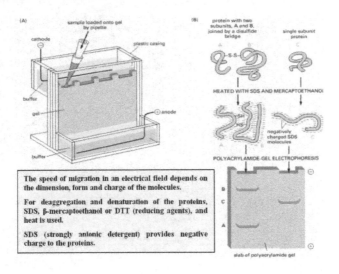

Figure 2. Polyacrylamide-gel electrophoresis (SDS-PAGE) used in proteome analysis [19].

2.1. Two-dimensional polyacrylamide gel electrophoresis (2D PAGE).

Two-dimensional polyacrylamide gel electrophoresis constitutes an analytical method capable of separating hundreds of proteins in a single analytical run. In this case, the gel, with the sample already applied, is submitted to an electrical field for two-dimensional separation. In the first dimension, separation occurs through isoelectric focalization, in which physical separation of the proteins occurs in terms of their respective isoelectric points on a strip of polyacrylamide with continuous gradation and known pH (IPG - immobilized pH gradient) submitted to increasing voltage. In the second dimension, the proteins under focus are submitted to polyacrylamide gel electrophoresis in the presence of SDS (SDS-PAGE) for separation according to their specific molecular masses (Figure 4). Thus, this is a technique that separates the proteins through different charges and masses.

The result of two-dimensional electrophoresis is a profile of spot distribution formed by single proteins or simple mixtures of proteins [21]. Each spot visualized in the gel may be considered as an orthogonal coordinate of a protein that migrated specifically in accordance with its isoelectric point (x axis) and its molecular mass (y axis), as shown in Figure 4.

The next step consists of staining the gel with silver, Coomassie blue, fluorescence, radioactive labeling or specific markers for phosphoproteins and glycoproteins, among others. This allows visualization of the protein expression pattern and photodocumentation of the gel

(Figure 5). After that, sectioning and digestion of selected spots of the gel are carried out and, finally, proteins of interest are identified by mass spectrometry integrated with a bioinformatics tool.

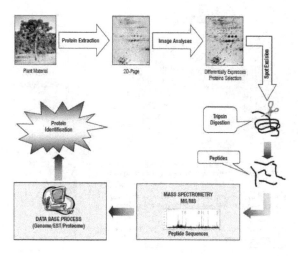

Figure 3. Stages of plant proteomics, using interface two-dimensional electrophoresis (2D-PAGE) and mass spectrometry [20].

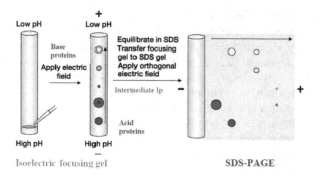

Figure 4. Two-dimensional electrophoresis 2D-PAGE used in analysis of proteomes [19].

Two-dimensional electrophoresis gels reflect the protein expression pattern of the biological sample analyzed and allow detection of variation of even a single amino acid between two isoforms or covalent modifications in the same protein thanks to change in the position of the spot.

It is important to highlight that each sample, depending on its nature, requires a specific type of processing for extraction and focalization. Therefore, it is expected that the user checks beforehand in related publications as to the protocols and methodologies that best suit the experimental needs.

Some limitations are associated with two-dimensional electrophoresis, such as low reproducibility and little power of automation. Nevertheless, reproducibility may be increased by defining optimal conditions for the electrophoresis, while automation of the process is only possible in relation to analysis of gels. Gel analysis software determines the spots and identifies those expressed differentially and their volumes, inferring a relative quantification of expression of that protein in comparison to the same spot of another gel [22]. Thus, by a process of subtraction, the differences among the different samples are revealed, as, for example, the presence, absence or intensity of the proteins. Thus, the proteins of interest may then be identified based on knowledge of the isoelectric point and of apparent molecular weight, determined by the two-dimensional gels [23].

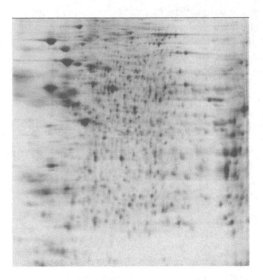

Figure 5. Proteins extracted and separated by two-dimensional (2D) gel electrophoresis and stained with Coomassie blue [24].

2.2. Differential in gel electrophoresis (DIGE).

An efficient procedure in the attempt to eliminate variation from gel to gel is use of the technique of differential in gel electrophoresis or DIGE (Figure 6), which allows analysis of up to three proteomes in a single gel. These results in one internal pattern common to all the gels and two different samples labeled with distinct fluorophores (CyDye) [25]. That way, only the proteins labeled with their own fluorophore are visualized. In addition, this technique uses labeling of proteins with a broad dynamic range of detection and has sensitivity greater than staining of the gels by silver methods, allowing proteomic studies of a quantitative nature to be performed with greater precision, accuracy and sensitivity [26].

2.3. Liquid chromatography

Another form used for separation of proteins is by means of liquid chromatography. The sample that is, for example, a mixture of peptides generated by proteolytic digestion from a protein extract passes through a first separation, by means of liquid chromatography, where the enriched peptide fractions are collected and applied in the spectrometer. As complete automation is the main target of the methods for large scale analyses, methods of separation were developed free of gel by reverse phase liquid chromatography connected with tandem mass spectrometry (LC/MS/MS). In Figure 7 the operational and equipment sequence involved in a typical analysis via LC/MS/MS is shown.

Greater automation is possible with multidimensional liquid chromatography, which uses different characteristics of the proteins in columns of distinct properties or in a single two-phase column [29]. The fraction eluted in the first column is directly introduced in the second column, which may be directly connected to the mass spectrometer. This technique, called MudPIT, is inserted in the context of the shotgun proteomic, in which greater resolution of the proteomes is possible, facilitating identification of the less abundant proteins frequently lost when gels are used [30].

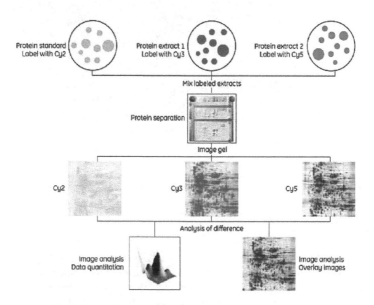

Figure 6. Differential in gel electrophoresis technique or DIGE [27].

Protein Identification with Chromatographic Separation (LC/MS/MS)

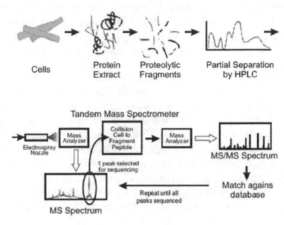

Figure 7. Protein identification with chromatographic separation (LC/MS/MS) [28].

2.4. Protein identification methods.

After separation of proteins, the next stage consists of their characterization and identification using mass spectrometry, which is a technique where the ratio between the mass and the charge (m/z) of ionized molecules in the gas phase is measured. In general, a mass spectrometer consists of an ionization source, a mass analyzer, a detector and a data acquisition system.

The great variety of spectrometers found on the market is the result of different combinations of types of sources of ionization and mass analyzers, which provide certain levels of sensitivity and accuracy in the results. At the ionization source, the molecules are ionized and transferred to the gas phase. In the mass analyzer, the ions formed are separated in accordance with their m/z ratios and later detected, usually by electron multiplier [31].

With the development of ever more specialized equipment for proteins, mass spectrometry has become a revolutionary tool in modern protein chemistry. This technology has allowed identification of proteins by a methodology called peptide mass fingerprinting. Rocha et al. (2003) [3], state that this methodology is based on protein digestion to be identified by a proteolytic enzyme, for example trypsin, producing fragments called peptides. The masses of these peptides obtained form a kind of fingerprinting of the protein, which are then determined with great acuity (0.1 to 0.5 Da) by mass spectrometry.

Special software allows comparing the peptide mass fingerprinting of the protein one wishes to identify with those theoretically generated for all the protein sequences present in the databases. If the protein sequence problem is in the database, it will immediately be identified [32].

2.5. Relative protein quantification

Large scale protein quantification methods make an estimate of relative expression possible by means of labeling with radioactive isotopes, fluorescents and light/heavy, allowing the same protein to be quantified in a relative way among differently labeled samples. Some of the most used radioactive isotopes are the iCAT (Isotopic coded affinity tag), iTRAQ (isobaric tags) and H_2O^{18}.

The iCAT consists of addition of a label that has affinity for cysteine residues and which has a bonded molecule of eight atoms of hydrogen or eight atoms of deuterium. One sample is labeled with the tag containing hydrogen and the other sample with the tag containing deuterium. After digestion of the proteins, the resulting peptides are identified by mass spectrometry. Equal peptides labeled in the two samples are identified by overlap of the peaks that show distinct m/z due to the type of bonded isotope, with the ratio between the area of the two peaks being a relative measure of the expression of that protein. According to Yi & Goodlett (2003) [33], the main problems associated with this technique are the need for the presence of cysteine residues, the high cost of the reagents and the greater time necessary for sequencing.

In the iTRAQ technique, labeling of proteins with tags and identification by mass spectrometry is also used. The tags bond to all the free amino groups at the N terminal of all the peptides and on the internal side chains with lysine residues and vary according to the reporter group they carry, and they may have 114, 115, 116 or 117Da, thus allowing for the quantification of proteins in up to four types of samples at the same time. The relative quantification is carried out in the same way as in the iCAT, but high cost has restricted its use [34].

The aforementioned techniques require the consumption of specific and expensive reagents. Nevertheless, the same goal may be achieved with a simpler labeling method in which the proteins are labeled with one or two atoms of O_2. These are incorporated in the carboxyl terminal by simply supplying a solution with H_2O for one sample and a solution with H_2O^{18} for the other sample. Thus, the relative abundance of the peptides that will differ by 2Da is estimated [35].

Another quantification technique is Stable isotope labeling by amino acids in cell culture, (SILAC) which, together with mass spectrometry and bioinformatics resources, has proven to be quite adequate in proteomic studies. It is a technique that detects differences in the abundance of proteins among cell cultures by means of isotopic labeling of proteins. Labeling with stable isotopes is obtained by supplying isotopically enriched amino acids to a cell culture and natural amino acids to the culture to be compared (Figure 8).

2.6. Analysis of post-translational modifications (PTM's).

Another area of great interest in plant proteomics is in regard to characterization of post-translational modifications or PTM's, essential for proteins to play their roles in the varied cell events, producing different proteins from the same gene.

These modifications occur at specific sites in the proteins [37] changing their physical, chemical and biological properties [38]. They may occur by means of cleavages or by the addition of a chemical group to one or more amino acids [39]. The main goals of PTM studies in pro-teomics are identifying the proteins that have them, mapping the sites where these modifi-cations occur, quantifying their occurrence at the different sites and characterizing cooperative PTM's [40].

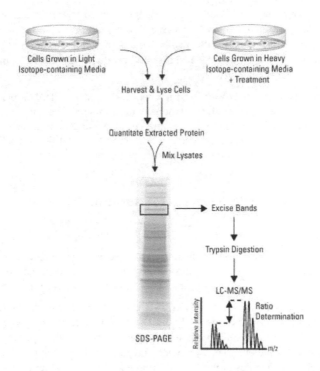

Figure 8. General outline of the SILAC technique [36].

The fact that covalent modifications result in changes in the protein molecular masses makes it possible for these modifications and the amino acids that carry them to be identified by mass spectrometry, allowing more than 300 different types of PTM's to be identified until now with the aid of this technique. Nevertheless, according to Mann and Jensen (2003) [41], mass spectrometry has reduced power of resolution of PTM's because they occur at low stoi-chiometric levels. This problem may be resolved by adopting fractioning methods prior to sequencing that allow enrichment of the sample for the proteins that have a certain type of PTM. Large scale modified protein enrichment systems are generally carried out by means of affinity chromatography.

One example is the IMAC system – a column of immobilization through affinity to a metal for isolation of phosphorylated proteins in which metal ions of Fe(III) are joined to the matrix to promote the isolation of proteins that have phosphorylate residues since the Fe(III) ion is capable of interacting in a reversible manner with the phosphate group of the modified peptide keeping it attached to the column [41].

Contrary to that which occurs with the reversible yet permanent PTM's, like glycosylation, low stoichiometry does not occur, but the addition of carbohydrates hinders the proteolytic digestion necessary for identification by mass spectrometry [21]. In addition, when the modified peptide is fragmented for sequencing, it loses sugar residues, impeding the identification of the modified amino acids. To resolve this problem, digestion of the proteins is performed so as to remove the sugar residues and produce a modification in the modified site that makes it identifiable [42].

Electrophoresis gels may also be used in enrichment of samples for PTM's as performed for detection of phosphorylations and glycosylations with commercially available kits. The modified proteins, specifically labeled in the gel, are visualized and excised for identification by mass spectrometry. One important aspect of the use of gels for identification of PTM is the possibility of visualizing the spots differentially expressed among samples that have the PTM.

3. Research dealing with proteomics in soybeans.

3.1. Food safety

In the case of food, proteins are especially important for evaluation of food safety because they may place consumer health at risk. That is because proteins may be involved in synthesis of toxins and antinutrients, as well as being a toxin, an antinutrient or even an allergenic [43].

Soybeans are an important source of food throughout the world, being consumed in daily meals of all types. It has also been widely used as a food substitute by people that have intolerance to lactose or other milk proteins [44]. Nevertheless, in this species are also found proteins considered allergenic. Thus, knowledge regarding the proteins with toxic/antinutritional potential present in this grain becomes fundamental for development of biotechnological strategies that would have the target of elimination or inactivation in the genome of these species of genes that codify for these proteins.

Therefore, application of proteomic analysis in this type of study has been widely discussed. In relation to products derived from genetically modified (GM) foods, proteomic techniques have been applied because they allow a wide-ranging approach and analysis of many variables simultaneously in the same sample [45]. Ocana et al. (2007) [46], studying GM proteins present in soybean and maize samples using proteomic analysis, identified the protein CP4 EPSPS, which confers tolerance to glyphosate herbicide. These samples were submitted to specific separation techniques followed by two-dimensional electrophoresis and mass spectrometry for detection and characterization of the proteome.

Related to allergies, various allergens belonging to the superfamily of cupins and prolamins have been identified in soybeans [47]. Research has suggested that a heterogeneous group of soybean proteins bond to the IgE antibody and are potential allergens as, for example, Gly m Bd 30k, β-conglycinin, Gly m Bd 28k, glycinin, Kunitz type protease inhibitor, some proteins present in the hull (Gly m 1.0101, Gly m 1.0102 e Gly m 2), profilin (Gly m 3), SAM 22 (Gly m 4), and other allergens like lectin and lipoxygenase [47,48]. According to Wilson et al. (2005) [49], in spite of the allergens identified in soybeans, the challenge of food researchers is developing a process for eradicating the immunodominant allergens, maintaining the functionality, nutritional value and effectiveness in the subsequent products derived from soybeans. For that reason, research has been developed using genetic engineering for silencing the soybean gene responsible for synthesis of the protein Gly m Bd 30K, one of the main soybean proteins that develop allergic reactions with serums of sensitive patients [44].

3.2. Biotic and abiotic factors

In a similar manner, various studies have shown that the proteomic approach is highly useful for investigation of crop response to environmental stresses because it compares the way the proteome is affected by different physiological conditions.

Saline stress is one of the many types of abiotic stresses that affect plants and compromise their yield. Salinity is a common agricultural problem in arid and semiarid regions and creates large unproductive areas. There has been an ever greater search for cultivars adaptable to this condition. Sobhanian et al. (2010) [10], used proteomic techniques to evaluate the metabolism of proteins in leaves, hypocotyls and roots submitted to different NaCl concentrations (Figure 9), thus leading to saline stress.

Results in soybeans suggest that, in adaptation to saline conditions, proteins perform different roles in each organ, and the proteins most affected by saline stress are those related to photosynthesis. Therefore, there is less energy production, and, consequently, reduction in plant growth. The conclusion suggests that the gene Glyceraldehyde-3-phosphate dehydrogenase may be, in the future, one of the target genes to improve tolerance to saline stress in this species.

Another type of abiotic stress studied in soybeans in which a proteomic approach is used is flooding stress [50,51]. Growing this species in areas subject to flooding makes the root environment anoxic, affecting nodulation or root growth. That way, plants respond with greater or less efficiency, allowing the distinction between cultivars which are tolerant and intolerant to this stress.

Proteomic analyses of soybean seedlings in response to flooding were undertaken by Shi et al. (2008) [52] to identify the key proteins involved in this process. To identify the first proteins produced in response to flooding, the roots of the seedlings were used for extraction of the proteins. The two-dimensional gel results suggest that cytosolic ascorbate peroxidase 2 (cAPX 2) is involved in response to flooding stress in young soybean seedlings.

In the case of drought stress, up-regulation of reactive oxygen species (ROS) scavengers such as superoxide dismutase (SOD) was reported in soybean seedlings [53]. The proteome analysis of two-day-old soybean seedlings subjected to drought stress by withholding of water for two days revealed a variety of responsive proteins involved in metabolism, disease/defense and energy including protease inhibitors [53]. The major reason for loss of crop yields under drought stress is a decrease in carbon gain through photosynthesis. Proteome analysis of soybean root under drought condition showed that two key enzymes involved in carbohydrate metabolism, UDP- glucose pyrophosphorylase and 2,3-bisphosphoglycerate independent phosphoglycerate mutase, were down-regulated upon exposure to drought [54]. The identification of proteins such as UDP-glucose pyrophosphorylase and 2,3- bisphosphoglycerate has provided new insights that may lead to a better understanding of the molecular basis of responses to drought stress in soybean

Figure 9. Soybean seedlings submitted to different concentrations of NaCl [10].

Stress by toxicity caused by the presence of high quantities of aluminum in the soil has also been investigated in soybeans from the perspective of proteomics [55,56]. Duressa et al. (2011) [56], studying cultivars tolerant and susceptible to high doses of aluminum, made proteomic analyses of roots, arriving at the conclusion that the greatest expression of enzymes involved with citrate synthesis would be a good strategy in the search for cultivars tolerant to this mineral (Figure 10).

Another focus of the study within the context of selection of superior soybean genotypes using the proteomic approach is exposure to ultraviolet radiation, which has gained importance with the prominent worldwide concern for global warming and the consequent degradation of the ozone layer. Xu et al. (2007) [57], studied the proteome of soybean leaves to investigate the protective role of flavonoids against the incidence of UV-B radiation. The authors suggest that high levels of flavonoid reduce the sensitivity of the plant to this radiation.

In relation to biotic stresses caused by pathogens like fungi, bacteria, nematodes and viruses, proteomic tools are also greatly used because they allow understanding of the plant-pathogen relationship [11,58,59,60] and also how the nodulation process occurs by means of symbiosis between the soybean roots and rhizobia [61]. In these cases, proteomic analysis provides the information that will be used by genetic breeding in the search for cultivars resistant to various diseases.

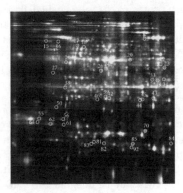

Figure 10. profile of aluminum regulated-proteins in PI 416937 72 h posttreatment [56].

Zhang et al. (2011) [58] evaluated the responses of cultivars tolerant and susceptible to the fungus *Phytophthora sojae* by means of two-dimensional electrophoresis. The authors observed 46 proteins being expressed (Figure 11), among which only 11% were related to plant defense.

In addition, proteomic studies that deal with seed development also play an essential role [62]. The data obtained may help to interpret the function of genes that determine protein concentration, considered as a key characteristic for genetic breeding of soybeans. Moreover, differential proteomic analyses designed to describe the changes that occur from maturation to senescence in organs and organelles have been reported. There is also already a soybean proteome database, providing information on the proteins involved in the soybean response to stress caused by drought, salinity and, principally, flooding [63].

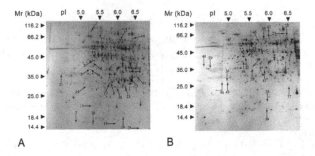

Figure 11. Identification of 26 and 20 protein spots from Yudou25 (A) and NG6255 (B), respectively. The numbers with arrows indicate the differentially expressed protein spots. Ip and Mr are shown on the gels [58].

4. Final considerations

In light of the above, proteomics in soybean studies contributes to diverse biotechnological applications, with its approach proving to be fundamental. Its use in the search for superior soybean materials has the purpose of comparing and contrasting genotypes for a determined type of stress and identifying the proteins that respond to the stress by means of changes in their levels of expression. The identification of these molecules and their respective functions will allow direction of breeding work, which should continue only with those that perform roles related to the characteristic of stress tolerance.

For that reason, it is essential to cross proteomic data with information also gathered from genomics, transcriptomics and metabolomics so as to check the correlation of the candidate proteins with the desired characteristic. The following stage aims to evaluate these proteins (genes) in regard to their segregation for the characteristic of interest or quantitative trait locus (QTL), that is, determine how much each one of them contributes to the characteristic of tolerance. Finally, the selected genes may be integrated in marker assisted selection (MAS) or in genetic transformation programs.

Author details

Cristiane Fortes Gris[1] and Alexana Baldoni[2]

1 Federal Institute of Southern Mines

2 Federal University of Lavras, Brazil

References

[1] Schmutz, J., Cannon, S. B., Schlueter, J., , J., Mitros, T., Nelson, W., Hyten, D. L., Song, Q., Thelen, J. J., Cheng, J., et al. (2010). Genome sequence of the paleopolyploid soybean. *Nature*, 463, 178-183, http://www.nature.com/nature/journal/463n7278/full/ nature08670.html, (accessing 20 february 2012).

[2] Pandey, A., & Mann, M. (2000). Proteomics to study genes and genome. Nature .http://65.199.186.23/nature/journal/ n6788/abs/405837a0.html (accessing 20 february 2012) , 405(6788), 837-846.

[3] Rocha, T. L., Costa, P. H. A., Magalhaes, J. C. C., Evaristo, R. G. S., Vasconcelos, E. A. R., Coutinho, M. V., Paes, N. S., Silva, M. C. M., & Rossi-de-Sa, M. F. (2005). Eletroforese bidimensional e análise de proteomas. *Comunicado Técnico. Embrapa Recursos Genéticos e Biotecnologia*, 136, 1-12.

[4] Mann, M. (2006). Functional and quantitative proteomics using SILAC. *Nature Review Molecular Cell Biology*, 7(12), 952-958, http://patf.lf1.cuni.cz/proteomika/silac_2006.pdf, (accessed 07 march 2012).

[5] Mcdonald, T., Sheng, S., Stanley, B., Cheng, D., Ko, Y., Cole, R. N., Pedersen, P., & Van Eyk, J. E. (2006). Expanding the subproteome of the inner mitochondria using protein preparation technologies. One- and two-dimensional liquid chromatography and two-dimensional gel electrophoresis. *Molecular & Cellular Proteomics*, 5(12), 2392-2411, http://www.mcponline.org/content/5/12/2392.full, (accessed 05 march 2012).

[6] Maor, R., Jones, A., Nuhse, T. S., Studholme, D. J., Peck, S. C., & Shirasu, K. (2007). Multidimensional protein identification technology (MudPIT) analysis of ubiquitinated proteins in plants. *Molecular & Cellular Proteomics*, 6, 601-610, http://www.mcponline.org/content/6/4/601.full, (accessed 2 march 2012).

[7] Natarajan, S., Xu, C., Caperna, T. J., & Garrett, W. M. (2005). Comparison of protein solubilization methods suitable for proteomic analysis of soybean seed proteins. *Analytical Biochemistry*, 342(2), 214-220, http://www.sciencedirect.com/science/article/pii/S0003269705003593, (accessed 2 march 2012).

[8] Krishnan, H. B., & Nelson, R. L. (2011). Proteomic analysis of high protein soybean (Glycine max) accessions demonstrates the contribution of novel glycinin subunits. *Journal of Agricultural and Food Chemistry*, 59, 2432-2439, http://pubs.acs.org/doi/abs/10.1021/jf104330n, (accessed 2 march 2012).

[9] Qureshi, M. I., Qadir, S., & Zolla, L. (2007). Proteomics-based dissection of stress response pathways in plants. *Journal of plant physiology*, 01, 1-13, http://www.sciencedirect.com/science/article/pii/S0176161707000995, (accessed 3 march 2012).

[10] Sobhanian, H., Razavizadeh, R., Nanjo, Y., Ehsanpour, A. A., Jazii, F. R., Motamed, N., & Komatsu, S. (2010). Proteome analysis of soybean leaves, hypocotyls and roots under salt stress. *Proteome Science*, 8, 19-33, http://www.biomedcentral.com/1477-5956/8/19, (accessed 26 february 2012).

[11] Liu, D., Chen, L., & Duan, Y. (2011). Differential proteomic analysis of the resistant soybean infected by soybean cyst nematode. *Heterodera glycines race 3. Journal of Agricultural Science*, 3(4), 160-167, http://www.ccsenet.org/journal/index.php/jas/article/view/9658, (accessed 26 february 2012).

[12] Castro, V. A. O. T., & Finardi, F. F. (2008). Análise Proteômica de amostras de soja GM e parental. I XXI Congresso Brasileiro de Ciência e Tecnologia de Alimentos e XV Seminário Latino Americano e do Caribe de Ciência e Tecnologia de Alimentos, Belo Horizonte, Brasil XXI SBCTA, 2008.

[13] Banks, R. E., Dunn, M. J., Hochstrasser, D. F., Sanchez, J., Blackstock, W., Pappin, D. J., & Selby, P. J. (2000). Proteomics: new perspectives, new biomedical opportunities. *The Lancet*, 356, 1749-1756, http://cmbi.bjmu.cn/cmbidata/proteome/reviews/02.pdf, (accessed 26 february 2012).

[14] Alfenas, A.C. (1998). Eletroforese de isoenzimas e proteínas afins: fundamentos e aplicações em plantas e microorganismos. *Viçosa, MG: UFV.*, 574.

[15] Tyers, M., & Mann, M. (2003). From genomics to proteomics. *Nature*, 422(6928), 193-197, http://www.ncbi.nlm.nih.gov/pubmed/12634792, (accessed 2 march 2012).

[16] Anderson, N. G., & Anderson, N. L. (1996). Twenty years of Two-dimensional electrophoresis: past, present and future. *Electrophoresis*, 17, 443-453, http://siba.unipv.it/fisica/articoli/E/Electrophoresis1996_17_443.pdf, (accessed 26 february 2012).

[17] Wilkins, M. R., Sanchez, J. C., Williams, K. L., & Hochstrasser, D. F. (1996). Current challenges and future applications for protein maps and posttranslational vector maps in proteome projects. *Electrophoresis*, 17(5), 830-838, http://onlinelibrary.wiley.com/doi/10.1002/elps.1150170504/abstract , (accessed 12 mai 2012).

[18] Wilkins, M. R., Willians, K. L., Appel, R. D., & Hochstrasser, D. (1997). *Proteome research: new frontiers in functional genomics.*, Germany, Springer-Verlag, 243.

[19] Stracham, T., & Read, A. P. (2004). *Human Molecular Genetics 3.*, New York, Garland Science.

[20] Balbuena, T. S., Dias, L. L. C., Martins, M. L. B., Chiquieri, T. B., Santa-Catarina, C., Floh, E. I. S., & Silveira, V. (2011). Challenges in proteome analyses of tropical plants. *Brazilian Journal of Plant Physiology*, 23(2), http://www.scielo.br/scielo.php?pid=S167704202011000200001 &script=sci_arttext, (accessed 09 march 2012).

[21] Pennington, S. R., & Dunn, M. J. (2001). *Proteomics: from protein sequence to function*, New York, Springer-Verlag e BIOS scientific Plubishers, 1v.

[22] López, J. L., Marina, A., Vásquez, J., & Alvarez, G. (2002). A proteomic approach to the study of the marine mussels. *Mytilus edulis and M. galloprovincialis. Marine Biology*, 141, 217-223, http://www.whoi.edu/science/B/people/mhahn/Lopez.pdf, (accessed 13 april 2012).

[23] James, P. (1997). Protein identification in the post-genome era: the rapid rise of proteomics. *Quaterly reviews of biophysics*, 30, 279-331.

[24] Applied Biomics: 2D Gel Staining. (2012). http://www.appliedbiomics.com/proteomics_differentstaining.html, (accessed 10may).

[25] Unlu, M., Morgan, M. E., & Minden, J. S. (1997). Difference gel electrophoresis: A single gel method for detecting changes in protein extracts. *Electrophoresis*, 18(11), 2071-2077, http://onlinelibrary.wiley.com/doi/10.1002/elps.1150181133/abstract, (accessed 15 april 2012).

[26] Tong, R., Shaw, J., Middleton, B., Rowlinson, R., Rayner, S., Young, J., Pognan, F., Hawkins, E., Currie, I., & Davison, M. (2001). Validation and development of fluorescent two-dimensional gel electrophoresis proteomics technology. *Proteomics*, 1, 377-396, http://www.ncbi.nlm.nih.gov/pubmed/11680884, (accessed 12 april 2012).

[27]   Healthcare, G.E. (2012). *Life Sciences 2D Electrophoresis Principles and Methods*, https://www.gelifesciences.com/gehcls_images/GELS/Related%20Content/Files/1335426794335/litdoc80642960_20120426103127.pdf, (accessed 10 april 2012).

[28]   Iowa State University. (2012). Q-Star Tandem Mass Spectrometry. http://www.protein.iastate.edu/q-star.html, (accessed 15 april).

[29]   Motoyama, A., & Yates, J. R. (2012). Multidimensional L.C. separations in shotgun proteomics. *Analytical Chemistry*, 80(19), 7187-7193, http://pubs.acs.org/doi/full/10.1021/ac8013669, (accessed 12 mai 2012).

[30]   Washburn, M. P., Wolters, D., & Yates, J. R. (2008). Large-scale analysis of the yeast proteome by multidimensional protein identification technology. *Nature Biotechnology*, 19, 242-247, http://proteome.gs.washington.edu/classes/Genome490/papers/Washburn_et_al_Nat_Biotech_2001.pdf, (accessed 28 april 2012).

[31]   Fenn, J. B., Mann, M., Meng, C. K., Wong, S. F., & Whitehouse, C. M. (1989). Electrospray ionization for mass spectrometry of large biomolecules. *Science*, 246, 64-71, http://www.sciencemag.org/content/246/4926/64.abstract, (accessed 25 february).

[32]   De Sousa, M. V., Fontes, W., & Ricart, C. A. O. (2003). Análise de Proteomas: O despertar da era pós-genômica. *Revista on line-Biotecnologia Ciência e Desenvolvimento*, 7, 12-14, http://www.biotecnologia.com.br/revista/bio07/analise.pdf, (accessed 10 april 2012).

[33]   Yi, E. C., & Goodlett, D. R. Quantitative protein profile comparisons using the isotope-coded affinity tag method. *Current Protocols in Protein Science*, 23, 2.1-23.2.11, http://onlinelibrary.wiley.com/doi/10.1002/0471140864.ps2302s34/abstract?systemMessage=Wiley+Online+Library+will+be+disrupted+on+9+June+from+10%3A00-12%3A00+BST+%2805%3A00-07%3A00+EDT%29+for+essential+maintenance, (accessed 28 april 2012).

[34]   Schmidt, C., & Urlaub, H. (2009). Itraq-labeling of in-gel digested proteins for relative quantification. *Methods in Molecular Biology*, 564(4), 207-226, http://www.springerlink.com/content/t864281256610gr4/#section=78222&page=1&locus=0, (accessed 28 april 2012).

[35]   Ye, X., Luke, B., Andresson, T., & Blonder, J. (2009). O stable isotope labeling in MS-based proteomics. *Briefings in Functional Genomics & Proteomics*, 8(2), 136-144, http://bfg.oxfordjournals.org/content/8/2/136.full, (accessed 15 april 2012).

[36]   Thermo Scientific: Pierce Protein Biology Products. (2012). SILAC Protein Quantitation Kits. http://www.piercenet.com/browse.cfm?fldID=01040418, (accessed 10 may).

[37]   Blom, N. (2004). Prediction of post-translational glycosylation and phosphorylation of proteins from the amino acid sequence. *Proteomics*, 4(6), 1633-1649, http://onlinelibrary.wiley.com/doi/10.1002/pmic.200300771/abstract , (accessed 15 april 2012).

[38]   Nesaty, V. J., & Suter, M. J. F. (2008). Analysis of environmental stress response on the proteome level. *Mass Spectrometry Reviews*, 27(6), 556-574, http://onlineli-

brary.wiley.com/doi/10.1002/mas.20177/abstract;jsessio-
nid=C9FCCA3D6997D990B13A7CAE38CB4F3E.d03t02 , (accessed 05 mai 2012).

[39] Mann, M., & Jensen, O. N. (2003). *Nature Biotechnology*, 21, 255-261, http://genom-
ics.wsu.edu/pages/teaching/MPS570%20PMB/ArticlesPDFs/MannandJensen.pdf, (accessed 02 april 2012).

[40] Seo, J., & Lee, K. J. (2004). Post-translational modifications and their biological functions: proteomic analysis and systematic approaches. *Journal of Biochemistry and Molecular   Biology,   37(1),   35-44,   http://bmbreports.org/jbmb/jbmb_files/
%5B37-1%5D0401271834_035-044.pdf, (accessed 12 april 2012).

[41] Vener, A. V., Harms, A., Sussman, M. R., & Vierstra, R. D. (2001). Mass spectrometric
resolution of reversible protein phosphorylation in photosynthetic membranes of
Arabidopsis thaliana. *Journal of Biological Chemistry*, 276(10), 6959-6966, http://
www.ufv.br/dbv/pgfvg/BVE684/htms/pdfs_revisao/genomica/protome_articles/Mass
%20Spectrometric%20Resolution%20of%20Reversible%20Protein%20Phosphoryla-
tion.pdf, (accessed 10 april 2012).

[42] Cantin, G. T., & Yates, J. R. (2004). Strategies for shotgun identification of post-translational modifications by mass spectrometry. *Journal of Chromatography A*, 1053(1-2),
7-14, http://www.sciencedirect.com/science/article/pii/S0021967304009719, (accessed
25 march 2012).

[43] Ruebelt, M. C., Leimgruber, N. K., Lipp, M., Reynolds, T. L., Nemeth, M. A., Ast-
wood, J. D., Engel, K. H., & Jany, K. D. (2006). Application of two-dimensional gel
electrophoresis to interrogate alterations in the proteome of genetically modified
crops. 1. Assessing analytical validation. *Journal of Agricultural Food Chem*, 54(6),
2154-2161, http://pubs.acs.org/doi/abs/10.1021/jf0523566 , (accessed 12 march 2012).

[44] Herman, E. (2003). Genetically modified soybeans and food allergies. *Journal of Experimental Botany*, 54(386), 1317-1319, http://www.ask-force.org/web/Soya/Herman-GM-
Soy-Allergies-2003.pdf, (acessed 12 march 2012).

[45] EFSA. GMO (2008). Panel Working Group on Animal Feeding Trials. Safety and nutritional assessment of GM plants and derived food and feed: The role of animal
feeding trials. Food and Chemical Toxicology (suppl. , 46(1), s2-s70.

[46] Ocana, M. F., Fraser, P. D., Patel, R. K., Halket, J. M., & Bramley, P. M. (2007). Mass
spectrometric detection of CP4EPSPS in genetically modified soya and maize. Rapid
Communication.   *Mass   Spectrometry,*   21(3),   319-328,   http://
onlinelibrary.wiley.com/doi/10.1002/rcm.2819/abstract , (accessed 13 march 2012).

[47] Boxtel, E.L.V. (2007). Protein quaternary structure and aggregation in relation to allergenicity. *Dissertation, Graduate School VLAG, chapter 1.*

[48] L'Hocine, L., & Boye, J. I. (2007). Allergenicity of soybean: New developments in
identification of allergenic proteins, cross-reactivities and hypoallergenization technologies. *Critical Reviews in Food Science and Nutrition*, 47(2), 127-143, http://

www.mendeley.com/research/allergenicity-soybean-new-developments-identifica-
tion-allergenic-proteins-crossreactivities-hypoallergenization-technologies/#,   (accessed 13 march 2012).

[49] Wilson, S., Blaschek, K., & De Mejia, E. G. (2005). Allergenic Proteins in Soybean:
Processing and Reduction of 34 Allergenicity. (accessed 10 march 2012). *Nutrition Reviews*,   63(2),   47-58,   http://onlinelibrary.wiley.com/doi/10.1111/j.
1753-4887.2005.tb00121.x/abstract.

[50] Komatsu, S., Kobayashi, Y., Nishizawa, K., Nanjo, Y., & Furukawa, K. (2010). Comparative proteomics analysis of differentially expressed proteins in soybean cell wall
during flooding stress. *Amino Acids*, 39(5), 1435-1449, http://www.springerlink.com/
content/e017718020408605/, (accessed 12 march 2012).

[51] Nanjo, Y., Maruyama, K., Yasue, H., Yamaguchi-Shinozaki, K., Shinozaki, K., & Komatsu, S. (2011). Transcriptional responses to flooding stress in roots including hypocotyl of soybean seedlings. Plant Molecular Biology http://www.springerlink.com/
content/j1783j7116046564/ (accessed 11 march 2012) , 77(1-2), 129-144.

[52] Shi, F., Yamamoto, R., Shimamura, S., Hiraga, S., Nakayama, N., Nakamura, T., Yukawa, K., Hachinohe, M., Matsumoto, H., & Komatsu, S. (2008). Cytosolic ascorbate
peroxidase 2 (cAPX 2) is involved in the soybean response to flooding. *Phytochemistry*,   69,   1295-1303,   http://www.sciencedirect.com/science/article/pii/
S0031942208000356, (accessed 10 mai 2012).

[53] Toorchi, M., Yukawa, K., Nouri, M.Z., & Komatsu, S. (2009). Proteomics approach for
identifying osmotic-stress-related proteins in soybean roots. *Peptides*, 30, 2108-2117,
http://www.sciencedirect.com/science/article/pii/S0196978109003647,   (accessed   09
mai 2012).

[54] Alam, I., Sharmin, S. A., Kim, K. H., Yang, J. K., Choi, M. S., & Lee, B. H. (2010). Proteome analysis of soybean roots subjected to short-term drought stress. *Plant and Soil*,
333(1-2), 491-505, http://www.springerlink.com/content/g9813jp3t5687604/, (accessed
12 april 2012).

[55] Zhen, Y., Qi, J. L., Wang, S. S., Su, J., Xu, G. H., Zhang, M. S., Miao, L., Peng, X. X.,
Tian, D., & Yang, Y. H. Comparative proteome analysis of differentially expressed
proteins induced by Al toxicity in soybean. *Physiologia Plantarum*, 131(4), 542-554,
http://onlinelibrary.wiley.com/doi/10.1111/j.13993054.2007.00979.x/abstract?system-
Message=Wiley+Online+Library+will+be+disrupted+on+9+June+from
+10%3A00-12%3A00+BST+%2805%3A00-07%3A00+EDT%29+for+essential+mainte-
nance&userIsAuthenticated=false&deniedAccessCustomisedMessage=, (accessed   10
march 2012).

[56] Duressa, D., Soliman, K., Taylor, R., & Senwo, Z. (2011). Proteomic Analysis of Soybean Roots under Aluminum Stress. *International Journal of Plant Genomics*, doi:
10.1155/2011/282531., http://www.hindawi.com/journals/ijpg/2011/282531/cta/, (accessed 10 march 2012).

[57] Xu, C., Sullivan, J. H., Garrett, W. M., Caperna, T. J., & Natarajan, S. (2008). Impact of solar Ultraviolet-B on the proteome in soybean lines differing in flavonoid contents. *Phytochemistry*, 69(1), 38-48, http://www.sciencedirect.com/science/article/pii/ S0031942207004025, (accessed 10 march 2012).

[58] Zhang, Y., Zhao, J., Xiang, Y., Bian, X., Zuo, Q., Shen, Q., Gai, J., & Xing, H. (2011). Proteomics study of changes in soybean lines resistant and sensitive to Phytophthora sojae. *Proteome Science*, 9(52), http://www.proteomesci.com/content/9/1/52, (accessed 26 february 2012).

[59] Mithofer, A., Muller, B., Wanner, G., & Eichacker, L. A. (2002). Identification of defence-related cell wall proteins in Phytophthora sojae-infected soybean roots by ESI-MS/MS. *Molecular Plant Pathology*, 3(3), 163-166, http://onlinelibrary.wiley.com/doi/ 10.1046/j.13643703.2002.00109.x/abstract;jsessio-nid=436F73B1434802655BD95CC22314118C.d01t04?deniedAccessCustomisedMes-sage=&userIsAuthenticated=false , (accessed 26 february 2012).

[60] Afzal, A. J., Natarajan, A., Saini, N., Iqbal, M. J., Geisler, M., El Shemy, H. A., Mungur, R., & Willmitzer, L. (2009). Lightfoot DA: The nematode resistance allele at the rhg1 locus alters the proteome and primary metabolism of soybean roots. *Plant Physiology*, 151(3), 1264-1280, http://www.ncbi.nlm.nih.gov/pmc/articles/PMC2773059/, (accessed 10 march 2012).

[61] Wan, J., Torres, M., Ganapathy, A., Thelen, J., Da, Gue. B., Mooney, B., Xu, D., & Stacey, G. (2005). Proteomic analysis of soybean root hairs after infection by Bradyrhizobium japonicum. *Molecular Plant-Microbe Interactions*, 18(5), 458-467, https:// mospace.umsystem.edu/xmlui/bitstream/handle/10355/3478/ProteomicAnalysisSoybeanRootHairs.pdf?sequence=1, (accessed 27 april 2012).

[62] Pandurangan, S., Pajak, A., Molnar, S. J., Cober, E. R., Dhaubhadel, S., Hernandez-Sebastia, C., Kaiser, W. M., Nelson, R. L., Huber, S. C., & Marsolais, F. (2012). Relationship between asparagine metabolism and protein concentration in soybean seed. *Journal of Experimental Botany*, 63(8), 3173-3184, http://jxb.oxfordjournals.org/content/ 63/8/3173, (accessed 11 march 2012).

[63] Sakata, K., Ohyanagi, H., Nobori, H., Nakamura, T., Hashiguchi, A., Nanjo, Y., Mikami, Y., Yunokawa, H., & Komatsu, S. (2009). Soybean Proteome Database: A data resource for plant differential omics. *Journal of Proteome Research*, 8(7), 3539-3548, http:// pubs.acs.org/doi/abs/10.1021/pr900229k , (accessed 11 march 2012).

Chapter 7

Use of Organelle Markers to Study Genetic Diversity in Soybean

Lidia Skuza, Ewa Filip and Izabela Szućko

Additional information is available at the end of the chapter

1. Introduction

Soybean is the most important crop provider of proteins and oil used in animal nutrition and for human consumption. Plant breeders continue to release improved cultivars with enhanced yield, disease resistance, and quality traits. It is also the most planted genetically modified crop. The narrow genetic base of current soybean cultivars may lack sufficient allelic diversity to counteract vulnerability to shifts in environmental variables. An investigation of genetic relatedness at a broad level may provide important information about the historical relationship among different genotypes. Such types of study are possible thanks to different markers application, based on variation of organelle DNA (mtDNA or cpDNA).

2. Mitochondrial genome

2.1. Genomes as markers

Typically, all sufficiently variable DNA regions can be used in genetic studies of populations and in interspecific studies. Because of in seed plants chloroplasts and mitochondria are mainly inherited uniparentally, organelle genomes are often used because they carry more information than nuclear markers, which are inherited biparentally. The main benefit is that there is only one allele per cell and per organism, and, consequently, no recombination between two alleles can occur. With different dispersal distances, genomes inherited biparentally, maternally and paternally, also reveal significant differences in their genetic variability among populations. In particular, maternally inherited markers show diversity within a population much better [1].

In gymnosperms the situation is somewhat different. Here, chloroplasts are inherited mainly paternally and are therefore transmitted through pollen and seeds, whereas mitochondria
are largely inherited maternally and are therefore transmitted only by seeds [2]. Since pollen
is distributed at far greater distances than seeds [3], mitochondrial markers show a greater
population diversity than chloroplast markers and therefore serve as important tools in con-
ducting genetic studies of gymnosperms [4]. Mitochondrial markers are also sometimes
used in conjunction with cpDNA markers [5].

Mitochondrial regions used in interspecific studies of plants, mainly gymnosperms, include, for example, introns of the NADH dehydrogenase gene *nad1* [4, 5, 6], the *nad7* intron 1 [7], the *nad5* intron 4 [3] and an internally transcribed spacer (ITS) of mitochondrial
ribosomal DNA [8, 9].

In addition to the aforementioned organelle markers, microsatellite markers [10, 11] and
simple sequence repeats (SSR) are often used in population biology, and sometimes also in
phylogeographic studies. Microsatellites are much less common in plants than in animals
[12]. However, they are present in both the nuclear genome and the organelle genome. Microsatellites may reveal a high variability, which may be useful in genetic studies of populations, whereas other sequences or methods such as fingerprinting do not detect mutations
sufficiently [9,10,13]. Inherited only uniparentally, organelle markers have a certain quality
in phylogeographic analyses. Since they are haploid, the effective population size should be
reduced after the analysis using these markers as compared to those in which nuclear markers are used [1, 14]. Smaller effective populations sizes should bring about faster turnover
rates for newly evolving genotypes, resulting in a clearer picture of past migration history
than those obtained using nuclear markers [15-17].

Initially, it was mainly in phylogeographic studies of animal species that mitochondrial
markers were used [18]. These studies have provided some interesting data on the begin-
nings and the evolutionary history of human population [19]. In contrast to studies of animals, using mitochondrial markers in studies of plants, especially angiosperms, is limited
[20]. Presently, cpDNA markers are most commonly used in phylogeographic studies of angiosperms, whereas mitochondrial markers are prevalent in studies of gymnosperms.

2.2. Plant mitochondrial DNA

Mitochondrial genomes of higher plants (208-2000 kbp) are much larger than those of vertebrates (16-17 kbp) or fungi (25-80 kbp) [21, 22]. In addition, there are clear differences in size
and organization of mitochondrial genomes between different species of plants. Intramolec-
ular recombination in mitochondria leads to complex reorganizations of genomes, and, in
consequence, to alternating arrangement of genes, even in individual plants, and the occurrence of duplications and deletions are common [23]. In addition, the nucleotide substitution
rate in plant mitochondria is rather low [24], causing only minor differences within certain
loci between individuals or even species. Extensively characterized circular animal mitochondrial genomes are highly conservative within a given species; they do not contain introns and have a very limited number of intergenic sequences [25]. Plant mitochondrial
DNA (mtDNA) contains introns in multiple genes and several additional genes undergoing

expression when compared to animal mitochondria, but most of the additional sequences in plants are not expressed and they do not seem to be esssentials [26]. The completely sequenced mitochondrial genomes are available for several higher plants, including *Arabidopsis thaliana* [27] or *Marchantia polymorpha* [28].

Restriction maps of nearly all plant mitochondrial genomes provide for the occurrence of the master circle with circular subgenomic molecules that arise after recombination among large direct repeats (> 1 kbp) [21, 29-36], which are present in most mitochondrial genomes of higher plants. However, such molecules, whose sizes can be predicted, are very rare or very difficult to observe. It can be explained by the fact that plant mitochondrial genomes are circularly permuted as in the phage T4 [37, 38]. Oldenburg and Bendich reported that mostly linear molecules in *Marchantia* mtDNA are circularly permuted with random ends [39]. It shows that plant mtDNA replication occurs similarly to the mechanism of recombination in the T4 [38].

Many reports that have appeared in recent years indicate that mitochondrial genome of yeasts and of higher plants exist mainly as linear and branched DNA molecules with variable size which is much smaller than the predicted size of the genomes [39-44]. Using pulsed field gel electrophoresis (PFGE) of in-gel lysed mitochondria from different species revealed that only about 6-12% of the molecules are circular [41, 44]. The observed branched molecules are very similar to the molecules seen in yeast in the intermediate stages of recombination of mtDNA [45] or the phage T4 DNA replication [37, 38].

In all but one known case (*Brassica hirta*) [46], plant mitochondrial genomes contain repeat recombinations. These sequences, ranging in length from several hundred to several thousand nucleotides (nt) exist at two different loci in the master circle, yet in four mtDNA sequence configurations [47]. These four configurations correspond to the reciprocal exchange of sequences 5' and 3' surrounding the repeat in the master circle, which suggests that the repeat mediates homologous recombination. Depending on the number and orientation of repeats, the master circle is a more or less complex set of subgenomic molecules [48].

Maternally inherited mutations, which are associated with mitochondria in higher plants, most often occur as a result of intra- and intergenic recombination. This happens in most cases of cytoplasmic male sterility (cms) [41, 49-51], in *chm*-induced mutation in *Arabidopsis* [52] and in non-chromosomal stripe mutations in maize [53]. In this way, it is assumed that the recombination activity explains the complexity of the variations detected in the mitochondrial genomes of higher plants.

2.3. Mitochondrial genome of soybean

The size of soybean mtDNA has been estimated to be approximately 400 kb [54-56]. Spherical molecules have also been observed by electron microscopy [55, 57].

Repeated sequences 9, 23 and 299 bp have been characterized in soybean mitochondria [58, 59]. Also, numerous reorganizations of genome sequences have been characterized among different cultivars of soybean. It has been demonstrated that they occur through homologous recombination produced by these repeat sequences [58, 60, 61], or through short elements that are part of 4.9kb PstI fragment of soybean mtDNA [62]. The 299 bp repeat sequence has been found in several copies of mtDNA of soybean and in several other higher plants, suggesting that this repeated sequence may represent a hot spot for recombination of mtDNA in many plant species [59, 62]. Previous results suggested that active homologous recombinations of mtDNA are present in at least some species of plants. Recently (2007) amitochondrial-targeted homolog of the *Escherichia coli recA* gene in *A. thaliana* has been identified [63]. However, the data on recombnation activity in plant mitochondria is still missing. The first data on such an activity in soybean was obtained in 2006 [64]. This discovery is supported by an analysis of mtDNA of soybean using electron microscopy and 2D-electrophoresis. The results suggest that only a small portion of mtDNA molecules undergoes recombination at any given time. Therefore the question is whether this recombination is essential to the functioning of mitochondria and to plant growth.

The repeated sequences of the *atp6*, *atp9* and *coxII* genes have been also characterized, but their recombination activity has not been analysed [65].

The first data for the restriction map of soybean mtDNA were obtained from the analysis of loci of the *atp4* gene [48]. In the vicinity of this gene two repeated sequences that show characteristics of recombination repeats have been found [47, 48]. Active recombination repeats were also identified in circular molecules smaller than 400 kb [55, 66]. These observations suggest that soybean mtDNA has multipartite structure that is similar to other plant mitochondrial genomes containing recombination repeats.

In the mitochondrial genome of cultivar Williams 82, recombinantly active repeats 1 kb and 2 kb have been described [48]. In a different repeat of 10 kb, surrounding both 1 kb and 2 kb repeats, two breakpoints have been identified. This recombination of smaller and larger repeats probably leads to the complex structure of genomes.

The analysis of restriction fragment length polymorphism (RFLP) of mtDNA seems to be a useful method in studying phylogenetic relationships within species.

Grabau et al. (1992) analyzed the genomes of 138 soybean cultivars [60]. Using 2.3 kb HindIII mtDNA probe from Williams 82 soybean cultivar revealed restriction fragment length polymorphisms (RFLPs), which allowed for the division of many soybean cultivars into four cytoplasmic groups: Bedford, Arksoy, Lincoln and soja-forage.

Subsequent analyses showed variations within, and adjacent to, the 4.8 kb repeats. Bedford cytoplasm turned out to be the only one that contains copies of the repeat in four different genomic environments, which indicates its recombination activity [61]. Lincoln and Arksoy cytoplasms contain two copies of the repeat and a unique fragment that appear to result from rare recombination events outside, but near, the repeat. In contrast, forage-soja cytoplasm contains no complete repeat, but it contains a unique truncated version of the repeat [61]. Sequence analysis revealed that truncating is caused by the recombination with a repeat of 9 bp CCCCTCCCC. The structural reorganization that occurred in the region around 4.8 kb repeat may provide a way to analyze the relationships between species and evolution within the soybean subgenus.

In order to determine the sources of cytoplasmic variability, Hanlon and Grabau (1995) studied the old cultivars of soybeans with the same 2.3-kb *Hind*III fragment and with a mtDNA fragment containing the *atp6* gene [62]. They showed that mtDNA RFLP analysis with these probes is useful for the classification of mitochondrial genomes of soybean. Grabau and Davies (1992) made a general classification of wild soybean using the 2.3-kb *Hind*III as a probe [68].

Mt type	Probe *coxI*			*coxII*			*atp6*		Reference
	Enzyme *Hind*III	*Bam*HI	*Eco*RI	*Hind*III	*Bam*HI	*Eco*RI	*Bam*HI	*Eco*RI	
Ic				1,6	5,8		5,0		[69]
Id				1,6	5,8		5,0;6,0 ;12,0		[69]
Ie				1,6	5,8		5,0; 12,0		[69]
Ik				1,6	5,8		5,0; 5,4; 5,8		[69]
IIg				1,3	7,0		1,0; 2,6		[69]
IIIa				1,2	8,5		2,4; 5,0		[69]
IIIb				1,2	8,5		2,9; 5,0		[69]
IIId				1,2	8,5		5,0;6,0; 12,0		[69]
IVa				3,5	8,1		2,4; 5,0		[69]
IVb				3,5	8,1		2,9; 5,0		[69]
IVc				3,5	8,1		5,0		[69]
IVf				3,5	8,1		2,4; 3,5; 5,0		[69]
IVh				3,5	8,1		2,6; 2,9		[69]
IVi				3,5	8,1		5,2; 12,0		[69]
Va				5,8	8,1		2,4; 5,0		[69]
V'j				5,8	15,0		5,0; 6,0		[69]
VIg				1,7	5,8		1,0; 2,6		[69]
VIIg				8,5	15,0		1,0; 2,6		[69]
mtI				1,6	5,8				[69]
mtII				1,3	7,0				[69]
mtIII				1,2	8,1				[69]
mtIV				3,5					[69]
mtV				5,8					[69]
mt-a							2,4; 5,0		[87]
mt-b							2,9; 5,0		[87]
mt-c							5,0		[87]
mt-d							5,0; 6,0; 12,0		[87]
mt-e							5,0; 12,0		[87]

Mt type	Probe	coxI			coxII			atp6		Reference
	Enzyme	HindIII	BamHI	EcoRI	HindIII	BamHI	EcoRI	BamHI	EcoRI	
mt-f								2,4; 3,5; 5,0		[87]
mt-g								1,0; 2,6		[87]
mt-h								2,6; 2,9		[87]
mt-m								2,9		[87]
mt-n								12,0		[87]
Ic		5,6	0,8; 2,5; 5,0	10,5	1,6	5,8	1,9	5,0	8,2; 12,0	[58]
Id		5,6	0,8; 2,5; 5,0	10,5	1,6	5,8	1,9	5,0; 6,0; 12,0	2,8; 6,0; 12,0	[58]
Ie		5,6	0,8; 2,5; 5,0	10,5	1,6	5,8	1,9	5,0; 12,0	2,8; 6,0; 12,0	[58]
Ik		5,6	0,8; 2,5; 5,0	10,5	1,6	5,8	1,9	5,0; 5,4; 5,8	2,8; 6,0; 12,0	[58]
IIg		8,5	0,8; 2,5; 5,0	9,0	1,3	7,0	4,8	1,0; 2,6	2,8; 3,0; 9,5	[58]
IIIb		5,6	0,8; 2,5; 5,0	10,5	1,2	8,5	6,2; 6,5	2,9; 5,0	6,0; 8,2; 12,0	[58]
IIId		5,6	0,8; 2,5; 5,0	10,5	1,2	8,5	6,2; 6,5	5,0; 6,0; 12,0	3,2; 6,2; 12,0	[58]
Iva		5,6	0,8; 2,5; 5,0	10,5	3,5	8,1	5,0	2,4; 5,0	3,0; 6,0; 12,0	[58]
IVb		5,6	0,8; 2,5; 5,0	10,5	3,5	5,8	5,0	2,9; 5,0	6,0; 8,2; 12,0	[58]
IVc		5,6	0,8; 2,5; 5,0	10,5	3,5	5,8	5,0	5,0	8,2; 12,0	[58]
IVf		5,6	0,8; 2,5; 5,0	10,5	3,5	5,8	5,0	2,4; 3,5; 5,0	3,2; 6,2; 12,0	[58]
IVh		5,6	0,8; 2,5; 5,0	10,5	3,5	5,8	5,0	2,6; 2,9	3,2; 6,2; 12,0	[58]
IVi		5,6	0,8; 2,5; 5,0	10,5	3,5	5,8	5,0	5,2; 12,0	3,2; 6,2; 12,0	[58]
Va		5,6	0,8; 2,5; 5,0	10,5	5,8	5,8	12,0	2,4; 5,0	3,0; 6,0; 12,0	[58]
Vb		5,6	0,8; 2,5; 5,0	10,5	5,8	5,8	12,0	2,9; 5,0	6,0; 8,2; 12,0	[58]
Vc		5,6	0,8; 2,5; 5,0	10,5	5,8	5,8	12,0	5,0	8,2; 12,0	[58]
V'j		5,6	0,8; 2,5; 5,0	10,5	5,8	15,0	1,6	5,0; 6,0	2,8; 6,0; 12,0	[58]

Mt type	Probe	coxI			coxII			atp6		Reference
	Enzyme	HindIII	BamHI	EcoRI	HindIII	BamHI	EcoRI	BamHI	EcoRI	
VIg		5,6; 8,5	0,8; 2,5; 5,2	5,0;9,0; 10,5	1,7	5,8	4,5	1,0; 2,6	2,8; 3,0; 4,3; 9,5; 12,0	[58]
VIIg		8,5	0,8; 5,0; 5,2	9,0	8,5	15,0	1,6	1,0; 2,6	2,8; 3,0; 9,5	[58]
VIIIc		5,6	0,8; 2,5; 5,0	10,5	8,5; 10,0	11,0; 15,0	1,6	5,0	8,2; 12,0	[58]
Combined chloroplast and mitochondrial genome type										
cpI +mtIIIb					1,2	8,5		2,9; 5,0		[89]
cpI +mtIVb					3,5	8,1		2,9; 5,0		[89]
cpI+mtIVc					3,5	8,1		5,0		[89]
cpII +mtIVb					3,5	8,1		2,9; 5,0		[89]
cpII +mtIVc					3,5	8,1		5,0		[89]
cpIII+mtIe					1,6	5,8		5,0; 12,0		[89]
cpIII +mtIVa					3,5	8,1		2,4; 5,0		[89]
cpIII +mtVIIIc					8,5; 10,0	11,0; 15,0		5,0		[89]

Table 1. Classification of mitochondrial genome types based on RFLPs using coxI, coxII and atp6 as probes. Sizes of hybridization signals (kb) are shown.

In their research Tozuka et al. (1998) used two fragments of mtDNA as probes: the 0.7-kb HindIII-NcoI fragment containing the coxII (the gene encoding the mitochondrial cytochrome oxidase subunit II) of wild soybean and the 0.66-kb StyI fragment containing the atp6 (the gene encoding the mitochondrial ATPase subunit 6) from Oenothera [69, 70] (Table 1).

Based on the RFLPs detected in gel-blot analysis with the coxII and atp6 probes, the harvested plants were divided into 18 groups. Five mtDNA types were described in 94% of the surveyed plants. The geographical distribution of mtDNA types revealed that in many regions soybean growing wild in Japan consisted of a mixture of plants with different types of mtDNA, sometimes even within a single location. Some of these mtDNA types have shown marked geographic clines among the regions. In addition, some wild soybeans had mtDNA types that were identical to those described in cultivated soybeans. These results suggest that mtDNA analysis could resolve maternal origin among of the genus Glycine subgenus Soja [69].

Kanazawa et al. (1998) gathered 1097 *G. soja* plants from all over Japan and analyzed their RFLP of mitochondrial DNA (mtDNA) using five probes (*coxI, coxII, atp6, atp9, atp1=atpA*) [58] (Table 1). 20 different types of mitochondrial genomes labeled as combinations of types I to VII and types from a to k were identified and characterized in this study. Nearly all the mtDNA types described for soybean cultivars also occurred in wild soybean.

The mitochondrial *atpA* gene was also analysed [48]. It was shown that in soybean this gene has a sequence in 90-97% identical with mitochondrial genes of other plants [71-81]. Sequence similarity is limited to the *atpA* coding region. An intriguing feature of the *atpA* open reading frame of soybean is an 642 nt overlap in the putative translation termination site onto an unidentified open reading frame of the *orf214*. The ends of the open reading frame contain four tandems of UGA codon that covers four tandems of AUG codon that initiates an unidentified *orf214* frame. The *atpA-orf214* region was found in soybean mtDNA in multiple sequence contexts. This can be attributed to the presence of two recombination repeats.

The open reading frame shares 79% of nucleotide identity with the *orf214* and is located in the same *atpA* locus position as in common bean *orf209* [82]. Since such organization is a repeat of overlapping the *atpB* and *atpE* reading frames in several chloroplast genes [83, 84], the probability that the *orf214* codes a different ATPase subunit cannot be evaluated because small ATPase subunits are poorly conserved [85].

So for a total of 26 mtDNA haplotypes of wild soybeans have been identified based on RFLP with probes from two mitochondrial genes: *cox2* and *atp6* [69, 86] (Table 1). The three most common haplotypes (Id, IVa and Va) are present in 43 populations. The distribution of mtDNA haplotypes varies among opulations [87]. Recently Shimamoto (2001) analyzed the genetic polymorphisms of mitochondrial genes subgenus *Soja* originating from China and Japan [88] (Table 1). As a result of these studies, 6 types of mitochondrial genomes were distinguished.

3. Chloroplast genome

As the result of the extensive research conducted in the past two decades, cpDNA analysis brought about fundamental changes to the systematics of plants. The chloroplast genome is ideal for phylogenetic analyses of plants for several reasons. First, it occurs abundantly in plant cells and is taxonomically ubiquitous. And since it is well researched, it can be easily tested in the laboratory conditions and analyzed in comparative programs. Moreover, it often contains marker structural features cladistically useful, and, above all, it exhibits moderate or low rate of nucleotide substitution [89]. In regard to the mitochondrial genome, and also to cpDNA, researchers use in their studies two distinct phylogenetic approaches [90], namely taxonomic checking of specific traits features of molecular cpDNA and sequencing of specific genes or regions.

3.1. Chloroplast genome of soybean

In estimating the phylogeny of plants belonging to *Glycine*, particular attention was paid to unusual and specific features of cpDNA. In the course of many studies on the variability of chloroplast genome, a breakthrough came in 1993, with a study on assessing phylogeny of seed plants. The study used a huge database of the nucleotide sequences of the *rbcL* gene [91], encoding the ribulose-1,5-bisphosphate carboxylase, large subunit. The accumulation of a number of comparative data on this chloroplast gene made it a frequent object of research. This is due to the fact that this gene's locus is large (> 1400 bp), and provides many phylogenetically informative traits. The rate of the *rbcL* evolution proved to be appropriate for assessing issues related to phylogeny of plants, especially on the medium and high taxonomic levels. Over the years other sequences from other species as well as many other genes with another chloroplast *atpB* gene coding H+ -ATPase subunits [92-95]. The *atpB-rbcL* sequence reaches different lengths in *Glycine* as well as in other seed plants. The study by Chiang (1998) shows that the size of the *atpB-rbcL* space in the studied species ranges from 524 bp to 1000 bp [5], where in the non-coding region the occurrence of deletions and insertions, as well as a number of nucleotide substitutions is a common phenomenon, which can also be observed in *Glycine*. In *Glycine max*, its chloroplast genome differs from the core set chloroplast DNA genes because of the presence of a single, large inversion of approximately 51 kb, in the area between the *rbcL* gene and the *rps16* intron [96]. This inversion is also present in other legumes: the mutation was reported in *Lotus* and *Medicago* [96]. In addition, the noncoding *atpB-rbcL* region is rich in AT, due to which most non-coding regions rich in these base pairs show a small number of functions [97, 98]. Therefore, this predisposes them for faster evolution, and hence for use in molecular systematics.

The summary phylogeny was based on sequence of several cpDNA genes from hundreds of spermatophytes including *Glycine* (Table 2). These genes can be divided into three classes. The genes encoding the photosynthetic apparatus structure form the first class. The second class includes the rRNA genes and genes encoding the chloroplast genetic apparatus. The last class consists of an average of about 30 tRNA encoding genes [99], although their number can vary from 20 to 40 [100, 101].

Genes	Products
Genes for the photosynthesis system	
rbcL	Ribulose -1,5- bisphosphate carboxylase, large subunit
psaA, B	Photosystem I, P700 apoproteins A1, A2
psaC	9kDa protein
psaA	Photosystem II, D1 protein
psaB	47kDa chlorophyll a-binding protein
psaC	43 kDa chlorophyll a-binding protein
psaD	D2 protein
psaE	Cytochrome b559 (8kDa protein)
psaF	Cytochrome b559 (4kDa protein)
psaH	10 kDa phosphoprotein

Genes	Products
Genes for the photosynthesis system	
psaI, J, K, L, M, N	–J, -K, -L, -M, -N-proteins
atpA, B, E	H$^+$-ATPase, CF$_1$ subunits α, β, ε
atp F, H, I	CF$_0$ subunits I, III, IV
petB, D	Cytochrome b$_6$ /f complex, subunit b$_6$, IV
nadA- K	NADH Dehydrogenase, subunits ND 1, NDI 1
Genes for the genetic system	
16S rRNA	16S rRNA
23S rRNA	23S rRNA
trnA -UGC	Alanine tRNA (UGC)
trnG- UCC	Gliycine tRNA (UCC)
rnH- GUG	Histidine tRNA (GUG)
trnI- GAU	Isoleucine tRNA (GAU)
trnK- UUU	Lysine tRNA (UUU)
trnL- UAA	Leucine tRNA (UAA)
rps2, 7, 12, 16	30S: ribosomal proteins CS2, CS7, CS12, CS16
rp12, 20, 32	50S: ribosomal proteins CL2, CL 20, CL32
rpoA, B, C1, C2	RNA polymerase, subunits α, β, β′, β″
matK	Maturase –like protein
sprA	Small plastid RNA
Others	
clpP	ATP-dependent protease, proteolytic subunit
irf168 (ycf3)	Intron- containing Reading frame (168 codons)

Table 2. Chloroplast genes for the photosynthesis system, for the genetic and others.

The complete size of the *Glycine max* chloroplast genome is 152,218 bp. It contains 25,574 bp of inverted repeats (IRa and IRb), which are separated by a unique small single copy (SSC) region (17,895 bp) [98]. In addition, this genome consists of a large single region (LSC) of unique sequences with 83,175 bp. The IR extends from the *rps19* gene up to the *ycf1*. The *Glycine* chloroplast genome contains 111 unique genes and 19 duplicate copies in the IR, amounting to a total of 130 genes. The cpDNA analysis has showed the presence of 30 different tRNAs in it and 7 of them are repeated within the IR regions. The genes are composed in 60% of encoding regions (52% are protein coding genes and 8% are RNA genes), and in 40% of non-coding regions, including both intergenic spacers and introns. The total content of GC and AT pairs in the *Glycine* chloroplast genes is 34% and 66% respectively. Distinctly higher percentage of AT pairs (70%) was observed in non-coding regions than in coding regions (62% AT) [98].

In comparison with other eukaryotic genomes, cpDNA is highly concentrated, for example, only 32% of the rice genome is non-coding. In *Glycine max* it is slightly more – 40%. Most of the non-coding DNA is found in very short fragments that separate functional genes. Some studies have shown complex patterns of mutational changes in the non-coding regions. Some of the best known regions in the chloroplast genome is the farther region of the *rbcL* gene in many legumes. This non-coding sequence is flanked by the *rbcL* and *psaI* (the gene encoding the polypeptide I of photosystem I).

3.2. Extent of IR in Glycine

Analysis of the IR (inverted repeats) regions in *Glycine max* has shown that they are separated by a large region and a small region of a unique sequence. In cpDNA repeated sequences are usually located asymmetrically, which results in the formation of long and short regions of a unique sequence [102]. The IR in *Glycine* is a region with 25,574 bp containing 19 genes. At the IR/LSC junction, at the ends of the 5' IR, there is the repeated *rps19* gene (68 bp), and at the junction of the IR/SSC and 5' ends the duplicated *ycf1* gene (478 bp) is located. In the course of study it was shown that comparing cpDNA IR region in *Medicago*, *Lotus*, *Glycine* and *Arabidopsis* indicates that there are changes within the IR in the two legumes. *Glycine* and *Lotus* have 478 bp and 514 bp of the *ycf1* duplicated, whereas *Arabidopsis* has 1,027 bp duplicated in the IR. This contraction of the IR in these legumes accounts for the smaller size of their IR and larger size of the SSC. In addition, contraction of the IR boundary in legumes, IRa has been lost in *Medicago*. This loss has resulted in *ndhF* (usually located in the SSC) being adjacent to *trnH* (usually the first gene in the LSC at the LSC/IRa junction). Loss of one copy of the IR in some legumes provides support for monophyly of six tribes [103-106]. Wolfe (1988) identified duplicated sequences of portions of two genes, 40 bp of *psbA* and 64 bp of *rbcL*, in the region of the IR deletion between *trnH* and *ndhF* in *Pisum sativum* and these duplications were later identified in broad bean (*Vicia faba*) [104,107]. According to many researchers, the IR region is considered the most conserved part of the chloroplast genome, and thus, it is responsible for stabilizing the plastid DNA molecules [108, 109]. Thus the loss of IR can be phylogenetically informative at the local level, as well as misleading at the global phylogeny level, because the IR loss likely occurred independently in more than one group of plants. Coniferous and some legumes (*Pisum sativum*, *Vicia faba*, *Medicago sativa*), for example, contain only one IR. Perhaps the lack of repeat sequences in these plants is associated with an increased incidence of rearrangement of chloroplast genomes [109].

Introns or intergenic sequences in legume chloroplast DNA have become extremely important tools in phylogenetic analyses aimed at systematizing of this species [110, 111]. Moreover, their microstructural changes occur with great frequency in the regions of cpDNA. The body of existing research suggests that mutations in the non-coding regions and relatively fast evolution of the organelle genome encoding regions can serve as valuable markers for the separation species in their evolutionary origin [110, 111]. The systematics of plants gen

erally considers chloroplast indeles to be phylogenetic markers, because of their low prevalence in comparison with nucleotide substitutions [5].

3.3. CpDNA markers

There are many methods of generating molecular markers that rely on site-specific amplification of a selected DNA fragment using polymerase chain reaction (PCR) and its further
processing (restriction analysis, sequencing). Initially the research on the plant genome
(mostly phylogenetic studies) used non-coding and coding sequences of chloroplast DNA.
With time, the genes or DNA segments located in the nuclear DNA, mitochondrial
(mtDNA) and chloroplast (cpDNA) found a prominent place among plant DNA markers.
Fully automated DNA sequencing made it possible to subject ever-newer regions of plant
DNA to comparative sequencing.

One of the most frequently sequenced cpDNA fragments in plant phylogeny of spermatophytes is the *rbcL* gene encoding a large ribulose bisphosphate carboxylase subunit (RUBIS-
CO), whose length in most plants is 1,428, 1,431 and 1,434 bp, and insertions and deletions
within it are extremely rare [94]. For many years this gene has been the subject of many
comprehensive phylogenetic analyses of subgenus *Glycine* [112-114]. The *rbcL* is most commonly used in the analyses at the family and genus levels, but there also exists research at
the lower levels, cultivars and wild soybean [98, 115, 116]. A marker with very similar characteristics to those of the *rbcL* (the rate of evolution, the length of 1497 bp) is a gene encoding
the ATP synthase β subunit – the *atpB* [94].The *matK* gene sequence, encoding maturase involved in splicing of the type II introns, and whose length is 1,550 bp is characterized by a
rapid rate of evolution that allows to use it in research at the species and genus levels [117,
118]. Frequent mutations in this gene make it unsuitable for studies at higher taxonomic levels. Other popular cpDNA sequences used in phylogenetic studies of legumes include the
ndhF (the gene encoding the NADH protein, which is a dehyd98rogenase subunit), 16S
rDNA, the non-coding *atpB-rbcL* region [94], or the *trnL* (UAA) intron and mediator between
the *trnL* (UAA) exon and the *trnF* (GAA) gene [96, 117- 119].

It should be noted that the rate of evolution for a specific DNA region to be used as a marker
can vary significantly not only among systematic groups, but also within these groups [98].
Moreover, each DNA fragment within the same group has a different rate of evolution, such
as the *ndhF* cpDNA sequence in the *Solanaceae* family, which provides about 1.5 times more
information in terms of parsimony than the *rbcL* [90]. Therefore each gene or any other DNA
fragment used as a genetic marker has a typical range of "taxonomic" or phylogenetic applications, which can vary significantly within a taxon. For this reason, the *rbcL* sequence has
been widely used in *Gycine* for many years at the species and genus levels [104, 117, 118].

3.4. The genetic diversity of soybeans

The importance of genetic variations in facilitating plant breeding and/or conservation strategies has long been recognized [121]. Molecular markers are useful tools for assaying genetic variation and provide an efficient means to link phenotypic and genotypic variation [122].

In recent years, the progress made in the development of DNA based marker systems has advanced our understanding of genetic resources. These molecular markers are classified as: (i) hybridization based markers i.e. restriction fragment length polymorphisms (RFLPs), (ii) PCR-based markers i.e. random amplification of polymorphic DNAs (RAPDs), amplified fragmentlength polymorphisms (AFLPs), inter simple sequence repeats (ISSRs) and micro-satellites or simple sequence repeats (SSRs), and (iii) sequence based markers i.e. single nucleotide polymorphisms (SNPs) [121, 123]. Majority of these molecular markers have been developed either from genomic DNA library (e.g. RFLPs or SSRs) or from random PCR amplification of genomic DNA (e.g. RAPDs) or both (e.g. AFLPs) [123]. Availability of an array of molecular marker techniques and their modifications led to comparative studies among them in many crops including soybean, wheat and barley [124-126]. Among all these, SSR markers have gained considerable importance in plant genetics and breeding owing to many desirable attributes including hypervariability, multiallelic nature, codominant inheritance, reproducibility, relative abundance, extensive genome coverage (including organellar genomes), chromosome specific location, amenability to automation and high throughput genotyping [127]. In contrast, RAPD assays are not sufficiently reproducible whereas RFLPs are not readily adaptable to high throughput sampling. AFLP is complicated as individual bands are often composed of multiple fragments mainly in large genome templates [123]. The general features of DNA markers are presented in Table 3.

	Molecular markers			
	EST–SSRs	SSRs	RFLPs	RAPDs/AFLPs/ISSRs
Need for sequence data	Essential	Essential	Not required	Not required
Level of polymorphism	Low	High	Low	Low-moderate
Dominance	Co-dominant	Co-dominant	Co-dominant	Co-dominant
Interspecific transferability	High	Low-moderate	Moderate-high	Low-moderate
Utility in Marker assisted selection	High	High	Moderate	Low-moderate
Cost and labour involved in generation	Low	High	High	Low-moderate

Table 3. Important features of different types of molecular markers.

The genetic diversity of wild and cultivated soybeans has been studied by various techniques including isozymes [128], RFLP [87], SSR markers [124], and cytoplasmic DNA markers [87, 128, 129]. Based on haplotype analysis of chloroplast DNA, cultivated soybean appears to have multiple origins from different wild soybean populations [129, 130].

Using PCR-RFLP method soybean chloroplast DNAs were classified into three main haplotype groups (I, II and III) [113, 130, 131]. Type I is mainly found in the species of cultivated soybean (*Glycine max*), while types II and III are often found in both the cultivated and wild forms of soybean (*Glycine soja*). Type III is by far the most dominant in the wild soybean species [113]. In *Glycine*, these types are widely used in evaluating cpDNA variability and in determining phylogenetic relationships between different types of cpDNA using different marker systems. According to Chen and Hebert (1999) [133] analysis of cpDNA sequence is not sufficient for when the analysis of population genetics, and so cpDNA polymorphism assessment methods must be constantly complemented with methods such as single-strand conformation polymorphism (SSCP) [134], or dideoxy fingerprinting (ddF) [135], and directed termination and polymerase chain reaction (DT-PCR). However, some researchers point out that there are many disadvantages of these methods, mainly because of their high cost and large amount of work necessary for obtaining the results. In their view, a single change in the regions of *Glycine* chloroplast DNA at the species and genus levels should be located on a local-specific markers, for example, non-coding regions, using PCR and sequencing.

Analyses of non-coding regions of cpDNA have been employed to elucidate phylogenetic relationship of different taxa [90]. Compared with coding regions, non-coding regions may provide more informative characters in phylogenetic studies at the species level because of their high variability due to the lack of functional constraints. Non-coding regions of cpDNA have been assayed either by direct sequencing [136-141], or by restriction-site analysis of PCR products (PCR-RFLP) [142-146]. In Small's opinion (1998) non-coding regions, which include introns and intergenic sequences, often show greater variability at nucleotides than at the encoding regions, which makes the non-coding regions good phylogenetic markers [139]. Mutations in the form of insertions and deletions are accumulated in noncoding regions at the same rate as nucleotide substitutions, and such kinds of mutations significantly accelerate changes in these regions. In many cases, insertions or deletions are related to short repeat sequences. Therefore, many researchers continually focus on the analysis of non-coding regions. Using RFLP method, Close et al. (1989) found six cpDNA haplotypes and described them in types, ranging from group I to VI, including cultivated and wild soybeans [147]. In the course of their research they found that groups I and II diverge from groups III to VI, thus dividing subgenus *Soja* into two main groups. They presented a hypothesis that group II can be distinguished form group III by two independent mutations. Similar groups of haplotypes in legumes were also obtained by Shimamoto et al, (1992) [128] and Kanazawa et al, (1998) [148], using a combination of *EcoRI* and *ClaI* RFLPs. In their classification, Kazanawa et al. (1998) relied on sequential analysis and found that differences in the three types described by Shimamoto et al. (1992) resulted from two single-base substitutions: one in the non-coding region, between the *rps11 and rpl36*, and the other in the 3' part of the coding region of the *rps3*. Based on the existing reports, Xu et al. (2000) sequenced nine non-coding regions of cpDNA for seven cultivars and 12 wild forms of soybean (*Glycine max, Glycine soja, Glycine tabacina, Glycine tomentella, Glycine microphylla, Glycine clandestina*) in order to verify earlier classification of *Glycine* [113]. In the course of their studies, they located eleven single-base changes (substitutions and deletions) in the collected 3849 database. They located five mutations in the distinguished haplotypes I and II, and seven mutations in type III. In addition, haplotypes I and II were identical and clearly different from the taxons in type III. This research has not yielded significant results, because different types of cpDNA could not originate monophyletically, but it contributed to finding a common ancestor in the course of evolution of *Glycine*. A neighbor joining tree resulting from the sequence data revealed that the subgenus *Soja* connected with *Glycine microphylla*, which formed a distinct clad from *Clycine clandestine* and the tetraploid cytotypes of *Glycine tabacina* and *Glycine tomentella*. Several informative length mutations of 54 to 202 bases, due to insertions or deletions, were also detected among the species of the genus *Glycine*.

3.5. Non-coding regions of the chloroplast genome as site-specific markers in Glycine

In the chloroplast genomes of legumes, including soybean, there are many non-coding regions, which are characterized by a faster rate of evolution when compared to the coding regions. As mentioned earlier some of the chloroplast genes have introns, yet their structure differs from those occuring in the nuclear genes, since in the case of cpDNA introns have a tendency to adopt secondary structure, which affects the model in which cpDNA introns evolve and it is enforeced by the secondary structure. This restriction in changes caused by mutations affects the functional requirements related to the formation of introns [98, 108]. As there are no adequate studies on the evolution of introns, it can be assumed that their evolution is similar to that of the protein-encoding genes. The loss of introns in the course of the evolution of chloroplast DNA is an interesting process. It has been discovered that *O. sativa* has 3 introns less in cpDNA than *M. polymorpha* and *N. tabacum*. The loss of an intron in the *rpl2* gene was researched in 340 species representing 109 families of angiosperms including *Glycine* [149]. When trying to determine the taxonomic position, the absence of this intron in a given gene shows that it was lost at least six times in the evolution of angiosperms. In *Glycine* 23 introns have been identified while in *Arabidopsis thaliana* there are 26 introns, mostly located in the same genes and in the same locations within those genes [98, 102].

Non-coding regions in chloroplast DNA have become a major source for phylogenetic studies within the species *Glycine* and in many other seed plants. Earlier, the most popular phylogeny sequences included encoding regions, such as the *rbcL* gene sequences that were designed to determine the phylogenetic relationships between species in major taxonomic groups [113, 136-141]. According to Taberlet et al. (1991) [119] the potential ability of non-coding regions of cpDNA was reserved for species located in the lower taxonomic levels while the non-coding regions, which include introns and intergenic sequences, often show greater variability at nucleotides than is evident in the coding regions, which predisposes them to be used in population studies involving *Glycine*, and others [139,142].

As the result, many studies on phylogenetic utility of non-coding regions have been published [110]. For example [150]: *trnH-psbA, trnS-tang;* [148]: *rps11-rpl36, rpl16-rps3,* [113]: *trnT-trnL.*

In cpDNA analysis of many plants, very conservative regions flanking areas with high variability are used. The more conservative regions, the higher the chance for the primers designed in the PCR reaction, which will be able to join the broader taxonomic group [96, 113]. The region occurring between the *trnT* (UGU) and the *trnF* (GAA) genes is a large single copy wich is suitable because of the conservativeness of the *trn* genes and several hundred base pairs of noncoding regions. The intergenic space between the *trnT* (UGU) and the *trnL* (UAA) 5' exon ranged from 298 bp to about 700 bp in the species studied by [119]. In the plant genomes completely sequenced by Sugiute, the length of this region is different and amounts to 770 bp in rice and 710 bp in tobacco. In *Marchantia polymorpha* it is 188 bp [151]. This region is located between the tRNA genes, just as the non-coding sequence located between the *trnL* (UAA) 3' exon and the *trnF* (GAA). Due to its catalytic properties and its secondary structure, the *trnL* (UAA) intron, which belongs to type I introns, is less variable and therefore of better utility for evolutionary studies at higher taxonomic levels [113]. Moreover, depending on the species, they show high frequency of insertions or deletions, which makes them potentially useful as genetic markers.

Region	Primer sequence (5 - 3)*	Annealing temperature	Reference
trnH–psbA	f:TGATCCACTTGGCTACATCCGCC	60°C	[99] (tobacco)
	r: GCTAACCTTGGTATGGAAGT		[150] (soybean)
trnS–trnG	f: GATTAGCAATCCGCCGCTTT	60°C	[99] (tobacco)
	r: TTACCACTAAACTATACCCGC		[99] (tobacco)
trnT–trnL	f: GGATTCGAACCGATGACCAT	60°C	[113] (soybean)
	r: TTAAGTCCGTAGCGTCTACC		[99] (tobacco)
trnL–trnF	f: TCGTGAGGGTTCAAGTCC	56°C	[99] (tobacco)
	r: AGATTTGAACTGGTGACACG		[99] (tobacco)
atpB–rbcL	f: GAAGTAGTAGGATTGATTCTC	58°C	[99] (tobacco)
	r: CAACACTTGCTTTAGTCTCTG		[99] (tobacco)
psbB–psbH	f: AGATGTTTTTGCTGGTATTGA	56°C	[99] (tobacco)
	r: TTCAACAGTTTGTGTAGCCA		[99] (tobacco)
rps11–rpl36	f:GTATGGATATATCCATTTCGTG	50°C	[148] (soybean)
	r: TGAATAACTTACCCATGAATC		[148] (soybean)
rpl16–rps3	f: ACTGAACAGGCGGGTACA	50°C	[148] (soybean)
	r: ATCCGAAGCGATGCGTTG		[148] (soybean)
ndhD–ndhE	f: GAAAATTAAGGAACCCGCAA	56°C	[99] (tobacco)
	r: TCAACTCGTATCAACCAATC		[99] (tobacco)

*f, forward primer; r, reverse primer

Table 4. Primers used for amplification of nine non-coding regions of soybean cpDNA.

In most studied species, the *trnL* (UAA) intron ranges in size from 254 - 767 bp. Its smaller fragment – the P6 loop – reaches a length of 10 - 143 bp. It is commonly applied in DNA barcoding. Its main limitation lies in its low homologousness with the species from the Gene Bank, which amounts to 67.3%, while the homologousness of the P6 loop is 19.5%. However, it also has some advantages: conservative primers projected form and trouble-free amplification process. Amplification of the P6 loop can be performed even in a very degraded DNA. The intron is well known and its sequences are used to determine phylogenetic relationships between closely related species or to identify a plant species [152]. The first universal primers for this region were designed more than 20 years ago [119]. However, it does not belong to the most variable non-coding regions in chloroplast DNA [108]. The *trnL* (UAA) intron is the only one belonging to group I introns in chloroplast DNA, which means that its secondary structure is highly conservative, with a possibility of changes in its conservative [113] and variability in regions [99, 153]. Consequently, comparing the diversity of the *trnL* intron sequences allows to obtain new primers that contain conservative regions and amplify short sections contained between them [152].

Thus, in angiosperms, using non-coding regions in research at lower levels of the genome is a routine practice [108]. A large number of non-coding regions of cpDNA has been located in angiosperms, some of which are highly variable, whereas others show relatively small variability [108]. In studying the chloroplast genome, many researchers looked for universal primers that would allow amplification of many non-coding regions of cpDNA (Table 4) [111, 113, 148, 150].

4. Conclusion

In phylogenetic and population studies of *Glycine*, genetic information contained not only in cpDNA but also in mtDNA are often analysed. Organelle DNA can be used to find species-specific molecular markers. Molecular markers are an important tool to systematize the species because their use allows for detecting the differences in the genes directly. The selection of appropriate sequences, which depends on the taxonomic level at which reconstruction of the origin is carried out is very important. The initial selection concerns non-conservative sequences, which are subject to fast evolution, because the more related the specimen are, the more changeable the region should be. The relatively slow rate of evolution of certain sequences may exclude statistically significant analyses within families or species, while the study of relationship between species, which phylogenetically are very distant, using more slowly evolving sequences can be very useful. Non-coding sequences show a faster rate of evolution than the coding sequences. These regions accumulate a greater number of insertion/deletion or substitution than the non-coding regions, and therefore may be more suitable for research at inter-or intra-genus levels.

Author details

Lidia Skuza*, Ewa Filip and Izabela Szućko

*Address all correspondence to: skuza@univ.szczecin.pl

Cell Biology Department, Faculty of Biology, University of Szczecin, Poland

References

[1] Petit, R. J., Duminil, J., Fineschi, S., Hampe, A., Salvini, D., & Vendramin, G. G. (2005). Comparative organization of chloroplast, mitochondrial and nuclear diversity in plant populations. *Molecular Ecology*, 14, 689-701.

[2] Wagner, D. B. (1992). Nuclear, chloroplast, and mitochondrial DNA polymorphisms as biochemical markers in population genetic analyses of forest trees. *New Forests*, 6, 373-390.

[3] Liepelt, S., Bialozyt, R., & Ziegenhagen, B. (2002). Wind-dispersed pollen mediates postglacial gene flow among refugia. USA. *Proceedings of the Natlional Academy of Sciences*, 99, 14590-14594.

[4] Johansen, A. D., & Latta, R. G. (2003). Mitochondrial haplotype distribution, seed dispersal and patterns of postglacial expansion of ponderosa pine. *Molecular Ecology*, 12, 293-298.

[5] Chiang, Y. C., Hung, K. H., Schaal, B. A., Ge, X. J., Hsu, T. W., & Chiang, T. Y. (2006). Contrasting phylogeographical patterns between mainland and island taxa of the Pinus luchuensis complex. *Molecular Ecology*, 15, 765-779.

[6] Jaramillo-Correa, J. P., Beaulieu, J., & Bousquet, J. (2004). Variation in mitochondrial DNA reveals multiple distant glacial refugia in black spruce (Picea mariana), a transcontinental North American conifer. *Molecular Ecology*, 13, 2735-2747.

[7] Godbout, J., Jaramillo, J. P., Beaulieu, J., & Bousquet, J. (2005). A mitochondrial DNA minisatellite reveals the postglacial history of jack pine (Pinus banksiana), a broad-range North American conifer. *Molecual Ecology*, 14, 3497-3512.

[8] Huang, S., Chiang, Y. C., Schaal, B. A., Chou, C. H., & Chiang, T. Y. (2001). Organelle DNA phylogeography of Cycas taitungensis, a relict species in Taiwan. *Molecular Ecology*, 10, 2669-2681.

[9] Pleines, T., Jakob, S. S., & Blattner, D. B. (2009). Application of non-coding DNA regions in intraspecific analyses. *Plant Systematic and Evolution*, 282, 281-294.

[10] Tautz, D. (1989). Hypervariability of simple sequences as a general source for polymorphic DNA markers. *Nucleic Acids Research*, 17, 6463-6471.

[11] Tautz, D., Trick, M., & Dover, G. A. (1986). Cryptic simplicity in DNA is a major source of genetic variation. *Nature*, 322, 652-656.

[12] Lagercrantz, U., Ellegren, H., & Andersson, L. (1993). The abundance of various polymorphic microsatellite motifs differs between plants and vertebrates. *Nucleic Acids Research*, 21, 1111-1115.

[13] Powell, W., Morgante, M., Mc Devitt, R., Vendramin, G. G., & Rafalski, J. A. (1995). Polymorphic simple sequence repeat regions in chloroplast genomes: applications to

the population genetics of pines. USA. *Proceedings of the Natlional Academy of Sciences,* 92, 7759-77763.

[14] Birky, C. W., Fuerst, P., & Maruyama, T. (1989). Organelle gene diversity under migration, mutation, and drift: equilibrium expectations, approach to equilibrium, effects of heteroplasmic cells, and comparison to nuclear genes. *Genetics,* 121, 613-627.

[15] Rendell, S., & Ennos, R. A. (2002). Chloroplast DNA diversity in Calluna vulgaris (heather) populations in Europe. *Molecular Ecology,* 11, 69-78.

[16] Hudson, R. R., & Coyne, J. A. (2002). Mathematical consequences of the genealogical species concept. *Evolution,* 56, 1557-1565.

[17] Kadereit, J. W., Arafeh, R., Somogyi, G., & Westberg, E. (2005). Terrestrial growth and marine dispersal? Comparative phylogeography of five coastal plant species at a European scale. *Taxon,* 54, 861-876.

[18] Avise, J. C. (2000). *Phylogeography: the history and formation of species,* Cambridge, Harvard University Press.

[19] Richards, M. B., Macauly, V. A., Bandelt, H. J., & Sykes, B. C. (1998). Phylogeography of mitochondrial DNA in western Europe. *Annals of Human Genetics,* 62, 241-260.

[20] Tomaru, N., Takahashi, M., Tsumura, Y., Takahashi, M., & Ohba, K. (1998). Intraspecific variation and phylogeographic patterns of Fagus crenata (Fagaceae) mitochondrial DNA. *American Journal of Botany,* 85, 629-636.

[21] Mackenzie, S., & Mc Intosh, L. (1999). Higher plant mitochondria. *Plant Cell,* 11, 571-585.

[22] Schuster, W., & Brennicke, A. (1994). The plant mitochondrial genome: physical structure, information content, RNA editing, and gene migration to the nucleus. *Annual Review of Plant Physiology and Plant Molecular Biology,* 45, 61-78.

[23] Palmer, J. D. (1992). Mitochondrial DNA in plant systematics: applicationsand limitations. *In: Soltis PS, Soltis DE, Doyle JJ. (ed.) Molecular systematics of plants,* New York, Chapman and Hall, 26-49.

[24] Wolfe, K. H., Li, W. H., & Sharp, P. M. (1987). Rates of nucleotide substitution vary greatly among plant mitochondrial, chloroplast, and nuclear DNAs. USA. *Proceedings of the Natlional Academy of Sciences,* 84, 9054-9058.

[25] Larsson, N. G., & Clayton, D. A. (1995). Molecular genetic aspects of human mitochondrial disease. *Annual Review of Genetics,* 29, 151-178.

[26] Binder, S., Marchfelder, A., & Brennicke, A. (1996). Regulation of gene expression in plant mitochondria. *Plant Molecular Biology,* 32, 303-314.

[27] Unseld, M., Marienfeld, J., Brandt, P., & Brennicke, A. (1997). The mitochondrial genome of Arabidopsis thaliana contains 57 genes in 366,924 nucleotides. *Nature Genetics,* 15, 57-61.

[28] Oda, K., Yamato, K., Ohata, E., Nakamura, Y., Takemura, M., Nozato, N., Akashi, K., Kanrgae, T., Ogura, Y., Kohchi, T., & Ohyama, K. (1992). Gene organization from the complete sequence of liverwort, Marchantia polymorpha, mitochondrial DNA: a primitive form of plant mitochondrial genome. *Journal of Molecular Biology*, 223, 1-7.

[29] Backert, S., Nielsen, B. L., & Borner, T. (1997). The mystery of the rings: structure and replication of mitochondrial genomes from higher plants. *Trends in Plant Science*, 2, 477-483.

[30] Andre, C. P., & Walbot, V. (1995). Pulsed-field gel mapping of maize mitochondrial chromosomes. *Molecular and General Genetics*, 247, 255-263.

[31] Bendich, A. J. (1993). Reaching for the ring: the study of mitochondrial genome structure. *Current Genetetics*, 24, 279-290.

[32] Fauron, C., Casper, M., Gao, Y., & Moore, B. (1995). The maize mitochondrial genome: dynamic, yet functional. *Trends in Genetics*, 11, 228-235.

[33] Fauron, C. M. R., Casper, M., Gesteland, R., & Albertsen, M. (1992). A multi-recombination model for the mtDNA rearrangements seen in maize cmsT regenerated plants. *The Plant Journal*, 2, 949-958.

[34] Palmer, J. D., & Herbon, L. A. (1988). Plant mitochondrial DNA evolves rapidly in structure, but slowly in sequence. *Journal of Molecular Evolution*, 28, 87-97.

[35] Palmer, J. D., & Shields, C. R. (1984). Tripartite structure of the Brassica campestris mitochondrial genome. *Nature*, 307, 437-440.

[36] Palmer, J. D., Makaroff, C. A., Apel, I. J., & Shirzadegan, M. (1990). Fluid structure of plant mitochondrial genomes: evolutionary and functional implications. *In: Clegg MT, O'Brien SJ (ed.) Molecular Evolution*, New York, Alan R. Liss, 85-96.

[37] Mosig, G. (1987). The essential role of recombination in phage T4 growth. *Annual Review of Genetics*, 21, 347-371.

[38] Mosig, G. (1998). Recombination and recombinationdependent DNA replication in bacteriophage T4. *Annual Review of Genetics*, 32, 379-413.

[39] Oldenburg, D. J., & Bendich, A. J. (2001). Mitochondrial DNA from the liverwort Marchantia polymorpha: circularly permuted linear molecules, head-to-tail concatemers, and a 50 protein. *Journal of Molecular Biology*, 310, 549-562.

[40] Backert, S., Dorfel, P., & Borner, T. (1995). Investigation of plant organellar DNA by pulsed-field gel electrophoresis. *Current Genetics*, 28, 390-399.

[41] Backert, S., Lurz, R., Oyarzabal, O. A., & Borner, T. (1997). High content, size and distribution of singlestranded DNA in the mitochondria of Chenopodium album (L.). *Plant Molecular Biology*, 33, 1037-1050.

[42] Bendich, A. J., & Smith, S. B. (1990). Moving pictures and pulsed-field gel electrophoresis show linear DNA molecules from chloroplasts and mitochondria. *Current Genetics*, 17, 421-425.

[43] Bendich, A.J. (1996). Structural analysis of mitochondrial DNA molecules from fungi and plants using moving pictures and pulsed-field gel electrophoresis. *Journal of Molecular Biology*, 225, 564-588.

[44] Oldenburg, D. J., & Bendich, A. J. (1996). Size and structure of replicating mitochondrial DNA in cultured tobacco cells. *Plant Cell*, 8, 447-461.

[45] Sena, E. P., Revet, B., & Moustacchi, E. (1986). In vivo homologous recombination intermediates of yeast mitochondrial DNA analyzed by electron microscopy. *Molecular and General Genetics*, 202, 421-428.

[46] Palmer, J. D., & Herbon, L. A. (1987). Unicircular structure of the Brassica hirta mitochondrial genome. *Current Genetics*, 11, 565-570.

[47] Stern, D. B., & Palmer, J. D. (1984). Recombination sequences in plant mitochondrial genomes: Diversity and homologies to known mitochondrial genes. *Nucleic Acid Research*, 12, 6141-6157.

[48] Chanut, F. A., Grabau, E. A., & Gesteland, R. F. (1993). Complex organization of the soybean mitochondrial genome: recombination repeats and multiple transcripts at the atpA loci. *Current Genetics*, 23, 234-247.

[49] Dewey, R. E., Levings, C. S., III., & Timothy, D. H. (1986). Novel recom- binations in the maize mitochondrial genome produce a unique transcriptional unit in the Texas male-sterile cytoplasm. *Cell*, 44, 439.

[50] Laver, H. K., Reynolds, S. J., Moneger, F., & Leaver, C. J. (1991). Mitochondrial genome organization and expression associated with cytoplasmic male sterility in sunflower (Helianthus annuus). *Plant Journal*, 1, 185-193.

[51] Johns, C., Lu, M., Lyznik, A., & Mackenzie, S. (1994). A mitochondrial DNA sequence is associated with abnormal pollen development in cytoplasmic male sterile bean plants. *Plant Cell*, 4, 435-449.

[52] Martinez-Zapater, J. M., Gil, P., Capel, J., & Somerville, C. R. (1992). Mutations at the Arabidopsis CHM locus promote rearrangements of the mitochondrial genome. *Plant Cell*, 4, 889-899.

[53] Newton, K., Knudsen, C., Gabay-Laughnan, S., & Laughnan, J. (1990). An abnormal growth mutant in maize has a defective mitochondrial cytochrome oxidase gene. *Plant Cell*, 2, 107-113.

[54] Levings, C. S., III., & Pring, D. R. (1979). Mitochondrial DNA of higher plants and genetic engineering. In: Setlow JK, Hollaender A (ed.). *Genetic engineering*, 1, New York, Plenum Press, 205-222.

[55] Bailey-Serres, J., Leroy, P., Jones, S. S., Wahleithner, J. A., & Wolstenholme, D. R. (1987). Size distribution of circular molecules in plant mitochondrial DNAs. *Current Genetics, 12,* 49-53.

[56] Grabau, E., Davis, W. H., & Gengenbach, B. G. (1989). Restriction fragment length polymorphism in a subclass of the mandarin soybean. *Crop Science, 29,* 1554-1559.

[57] Synenki, R. M., Levings, C. S., III., & Shah, D. M. (1978). Physicochemical characterization of mitochondrial DNA from soybean. *Plant Physiology, 61,* 460-464.

[58] Kanazawa, A., Tozuka, A., Kato, S., Mikami, T., Abe, J., & Shimamoto, Y. (1998). Small interspersed sequences that serve as recombination sites at the cox2 and atp6 loci in the mitochondrial genome of soybean are widely distributed in higher plants. *Current Genetics, 33,* 188-198.

[59] Kato, S., Kanazawa, A., Mikami, T., & Shimamoto, Y. (1998). Evolutionary changes in the structures of the cox2 and atp6 loci in the mitochondrial genome of soybean involving recombination across small interspersed sequences. *Current Genetics, 34,* 303-312.

[60] Grabau, E. A., Davis, W. H., Phelps, N. D., & Gengenbach, B. G. (1992). Classification of soybean cultivars based on mitochondrial DNA restriction fragment length polymorphisms. *Crop Science, 32,* 271-274.

[61] Moeykens, C. A., Mackenzie, S. A., & Shoemaker, R. C. (1995). Mitochondrial genome diversity in soybean: repeats and rearrangements. *Plant Molecular Biology, 29,* 245-254.

[62] Hanlon, R. W., & Grabau, E. A. (1997). Comparison of mitochondrial organization of four soybean cytoplasmic types by restriction mapping. *Soybean Genetics Newsletter, 24,* 208-210.

[63] Khazi, F. R., Edmondson, A. C., & Nielsen, B. L. (2003). An Arabidopsis homologue of bacterial RecA that complements an E. coli recA deletion is targeted to plant mitochondria. *Molecular and General Genetics, 269,* 454-463.

[64] Manchekar, M., Scissum-Gunn, K., Song, D., Khazi, F., Mc Lean, S. L., & Nielsen, B. L. (2006). DNA Recombination Activity in Soybean Mitochondria. *Journal of Molecular Bioliology, 356,* 288-299.

[65] Grabau, E. A., Havlik, M., & Gesteland, R. F. (1988). Chimeric Organization of Two Genes for the Soybean Mitochondrial Atpase Subunit 6. *Current Genetics, 13,* 83-89.

[66] Wissinger, B., Brennicke, A., & Schuster, W. (1992). Regenerating good sense: RNA editing and trans-splicing in plant mitochondria. *Trends in Genetics, 8,* 322-328.

[67] Hanlon, R., & Grabau, E. A. (1995). Cytoplasmic diversity in old domestic varieties of soybean using two mitochondrial markers. *Crop Science, 35,* 1148-1151.

[68] Grabau, E. A., & Davis, W. H. (1992). Cytoplasmic diversity in Glycine soja. *Soybean Genetics Newslett, 19,* 140-144.

[69] Tozuka, A., Fukushi, H., Hirata, T., Ohara, M., Kanazawa, A., Mikami, T., Abe, J., & Shimamoto, Y. (1998). Composite and clinal distribution of Glycine soja in Japan revealed by RFLP analysis of mitochondrial. *DNA Theoretical and Applied Genetics*, 96, 170-176.

[70] Schuster, W., & Brennicke, A. (1987). Nucleotide sequence of the Oenothera ATPase subunit 6 gene. Nucleic Acids Research ., 15, 9092.

[71] Braun, C. J., & Levings, C. S., III. (1985). Nucleotide Sequence of the F(1)-ATPase alpha Subunit Gene from Maize Mitochondria. *Plant Physiology*, 79(2), 571-577.

[72] Isaac, G. I., Brennicke, A., Dunbar, M., & Leaver, C. J. (1985). The mitochondrial genome of fertile maize (Zea mays L.) contains two copies of the gene encoding the alpha-subunit of the F1-ATPase. *Current Genetics*, 10(4), 321-328.

[73] Schuster, W., & Brennicke, A. (1986). Pseudocopies of the ATPase a-subunit gene in Oenothera mitochondria are present on different circular molecules. *Molecular and General Genetics*, 204, 29-35.

[74] Morikami, A., & Nakamura, K. (1987). *Structure and expression of pea mitochondrial F1ATPase alpha-subunit gene and its pseudogene involved in homologous recombination, J Biochem*, 101, 967-976.

[75] Chaumont, F., Boutry, M., Briquet, M., & Vassarotti, A. (1988). Sequence of the gene encoding the mitochondrial F1ATPase alpha subunit from Nicotiana plumbaginifolia. *Nucleic Acids Research*, 16(13), 6247.

[76] Schulte, E., Staubach, S., Laser, B., & Kück, U. (1989). Wheat mitochondrial DNA: organization and sequences of the atpA and atp9 genes. *Nucleic Acids Research*, 17(18), 7531.

[77] Kadowaki, K., Kazama, S., & Suzuki, S. (1990). Nucleotide sequence of the F_1-ATPase a subunit genes from rice mitochondria. *Nucleic Acids Research*, 18, 1302.

[78] Köhler, R. H., Lössl, A., & Zetsche, K. (1990). Nucleotide sequence of the F1ATPase alpha subunit gene of sunflower mitochondria. *Nucleic Acids Research*, 18, 4588.

[79] Makaroff, C. A., Apel, I. J., & Palmer, J. D. (1990). Characterization of radish mitochondrial atpA: influence of nuclear background on transcription of atpA-associated sequences and relationship with male sterility. *Plant Molecular Biology*, 15(5), 735-746.

[80] Siculella, L., D'Ambrosio, L., De Tuglie, A. D., & gauerani, R. (1990). Minor differences in the primary structures of atpa genes coded on the mt DNA of fertile and male sterile sunflower lines. *Nucleic Acids Research*, 18(15), 4599.

[81] Handa, H., & Nakajima, K. (1991). Nucleotide sequence and transcription analyses of the rapeseed (Brassica napus L.) mitochondrial F1ATPase alpha-subunit gene. *Plant Molecular Biology*, 16, 361-364.

[82] Chase, C. D., & Ortega, V. M. (1992). Organization of ATPA coding and 3' flanking sequences associated with cytoplasmic male sterility in Phaseolus vulgaris. L. *Current Genetics, 22*, 147-153.

[83] Zurawski, G., Bottomley, W., & Whitfeld, P. R. (1982). Structures of the genes for the beta and c subunits of spinach chloroplast ATPase indicates a dicistronic mRNA and an overlapping translation stop/start signal. USA. *Proceedings of the Natlional Academy of Sciences, 79*, 6260-6264.

[84] Krebbers, E. T., Larrinua, I. M., Mc Intosh, L., & Bogorad, L. (1982). The maize chloroplast genes for the beta and epsilon subunits of the photosynthetic coupling factor CF1 are fused. *Nucleic Acids Research, 10*(16), 4985-5002.

[85] Walker, J. E., Fearnley, I. M., Gay, N. J., Gibson, B. W., Northrop, F. D., Powell, S. J., Runswick, M. J., Saraste, M., & Tybulewicz, V. L. J. (1985). Primary structure and subunit stoichiometry of F1-ATPase from bovine mitochondria. *Journal of Molecular Biology, 184*(4), 677-701.

[86] Shimamoto, Y., Fukushi, H., Abe, J., Kanazawa, A., Gai, J., Gao, Z., & Xu, D. (1998). RFLPs of chloroplast and mitochondrial DNA in wild soybean, Glycine soja, growing in China. *Genetic Resources and Crop Evolution, 45*, 433-439.

[87] Abe, J. (1999, 13th-15th October). The genetic structure of natural populations of wild soybeans revealed by isozymes and RFLP's of mitochondrial DNS's: possible influence of seed dispersal, cross pollination and demography. Japan. *In: The Seventh Ministry of Agriculture, Forestry and Fisheries (MAFF), Japan International Workshop on Genetic Resources Part I Wild Legumes*, National Institute of Agrobiological Resources Tsukuba, Ibaraki, 143-159.

[88] Shimamoto, Y. (2001). Polymorphism and phylogeny of soybean based on chloroplast and mitochondrial DNA analysis. *Japan Agricultural Research Quarterly (JARQ), 35*(2), 79-84.

[89] Cleeg, M. T., & Zurawski, G. (1992). Chloroplast DNA and the study of plant phylogeny. New York, *In: Soltis PS, Soltis DE, Doyle JJ (eds.) Present status and future procpects in Molecular Systematics of Plants*, Chapman &Hall, 1-13.

[90] Olmstead, R. G., & Palmer, J. D. (1994). Chloroplast DNA systematics: a review of method and data analysis. *American Journal of Botany, 81*, 1205-1224.

[91] Chase, M., & 41others, . (1993). Phylogenetics of seed plants: An analysis of nucleotide sequences from the plastid gene rbcL. *Annals Missouri Botany Gardens, 80*, 528-580.

[92] Parkinson, C. L., Adams, L., & Palmer, D. (1999). Multigene analyses identify the three earliest lineages of extant flowering plants. *Current Biology, 9*, 1485-1488.

[93] Qui, Y. L. (1999). The earliest angiosperms: Evidence from mitochondrial, plastid, and nuclear genomes. *Nature, 402*, 404-407.

[94] Soltis, D. E., Soltis, P. S., & Doyle, J. J. (1998). Molecular systematic of Plants II. *DNA sequencing*, Boston MA, Kluwer.

[95] Soltis, P. S., Soltis, D. E., & Chase, M. W. (1999). Angiosperm phylogeny inferred from multiple genes as a tool for comparative biology. *Nature*, 402, 402-404.

[96] Kato, T., Kaneko, T., Sato, S., Nakamura, Y., & Tabata, S. (2000). Complete structure of the chloroplast genome of a legume, Lotus japonicus. *DNA Research*, 7, 323-330.

[97] Li, W. H. (1997). *Molecular Evolution*, Sunderland, Sinauer Associates, 487.

[98] Saski, C.h., Lee, S. B., Daniell, H., Wood, T. C., Tomkins, J., Kim, H. G., & Jansen, R. K. (2005). Complete chloroplast genome sequence of Glycine max and comparative analyses with other legume genomes. *Plant Molecular Biology*, 59, 309-322.

[99] Shinozaki, K., Ohme, M., Tanaka, M., Wakasugi, T., Hayashida, N., Matsubayashi, T., Zaita, N., Chunwongse, J., Obokata, J., Yamaguchi-Shinozaki, K., Ohto, C., Torazawa, K., Yamada, K., Kusuda, J., Takaiwa, F., Kato, A., Tohdoh, N., Shimada, H., & Sugiura, M. (1986). The complete nucleotide sequence of tobacco chloroplast genome: its gene organization and expression. *EMBO Journal*, 5, 2043-2049.

[100] Sugiura, M., & Shimada, H. (1991). Fine- structural features of the chloroplast genome- comparison of the sequenced chloroplast genomes. *Nucleic Acids Research*, 19, 983-995.

[101] Sugiura, M., & Sugita, M. (1992). The chloroplast genome. *Plant Molecular Biology*, 19, 149-168.

[102] Woźny, A. (1990). *Wykłady i ćwiczenia z biologii komórki*, Warszawa, Wydawnictwo Naukowe PWN.

[103] Palmer, J. D. (1985). Evolution of chloroplast and mitochondrial DNA in plants and algae. *In: MacIntyre RJ. (ed.) Monographs in Evolutionary Biology*, New York:, Molecular Evolutionary Genetics Plenum Press, 131-240.

[104] Wolfe, K.H. (1988). The site of deletion of the inverted repeat is pea chloroplast DNA contains duplicated gene fragments. *Current Genetic*, 13, 97-99.

[105] Palmer, J. D., Osorio, B., Aldrich, J., & Thompson, W. F. (1987). Chloroplast DNA evolution among legumes: loss of a large inverted repeat occurred prior to other sequence rearrangements. *Current Genetic*, 11, 275-286.

[106] Lavin, M., Doyle, J. J., & Palmer, J. D. (1990). Evolutionary significance of the loss of the chloroplast-DNA inverted repeat in the Leguminosae subfamily Papilionoideae. *Evolution*, 44, 390-402.

[107] Herdenberger, F., Pillay, D. T. N., & Steinmetz, A. (1990). Sequence of the trnH gene and the inverted repeat structure deletion of the broad bean chloroplast genome. *Nucleic Acids Research*, 18, 1297.

[108] Shaw, J., Lickey, E. B., Beck, J. T., Farmer, S. B., Liu, W., Miller, J., Siripun, K. C., Winder, C. T., Schilling, E. E., & Small, R. L. (2005). The tortoise and the here II: rela-

tive utility of 21 noncoding chloroplast DNA sequences for phylogenetic analysis. *American Journal of Botany*, 92, 142-166.

[109] Ravi, V., Khurana, J. P., Tyagi, A. K., & Khurana, P. (2008). An update on chloroplast genomes. *Plant Systematics and Evolution*, 271101-122.

[110] Kelchner, S. A. (2000). The evolution of non-coding chloroplast DNA and its application in plant systematics. *Annals of the Missouri Botanical Garden*, 87, 482-498.

[111] Perry, A. S., Brennan, S., Murphy, D. J., & Wolfe, K. H. (2006). Evolutionary re-organisation of a large operon in Adzuki bean chloroplast DNA caused by inverted repeat movement. *DNA Research*, 9, 157-162.

[112] Close, P. S., Shoemaker, R. C., & Keim, P. (1989). Distribution of restriction site polymorphism within the chloroplast genome of the genus Glycine, subgenus Soja. *Theoretical and Applied Genetics*, 77, 768-776.

[113] Xu, D. H., Abe, J., Kanazawa, A., Sakai, M., & Shimamoto, Y. (2000). Sequence variation of non-coding regions of chloroplast DNA of soybean and related wild species and its implications for the evolution of different chloroplast haplotypes. *TAG Theoretical and Applied Genetics*, 101, 5-6, 724-732.

[114] Kajita, T., Ohashi, H., Tateishi, Y., Bailey, C. D., & Doyle, J. J. (2001). rbcL and legume phylogeny, with particular reference to Phaseoleae, Millettieae, and allies. *Systematic Botany*, 26, 515-536.

[115] Nickret, D. L., & Patrick, J. A. (1998). The nuclear ribosomal DNA intergenic spacer of wild and cultivated soybean have low variation and cryptic subrepeats. *NRC Canada. Genome*, 41183-192.

[116] Palmer, J. D., Adams, K. L., Cho, Y., Parkinson, C. L., Qiu, Y. L., & Song, K. (2000). Dynamic evolution of plant mitochondrial genomes: mobile genes and introns and highly variable mutation rates. U.S.A., *Proceedings of the National Academy of Sciences*, 97, 6960-6966.

[117] Pennington, R. T., Klitgaard, B. B., Ireland, H., & Lavin, M. (2000). New insights into floral evolution of basal Papilionoideae from molecular phylogenies. *In: PS Herendeen A Bruneau (ed.) Advances in Legume Systematics Kew*, UK, 9, 233-248.

[118] Wojciechowski, M. F., Lavin, M., & Sanderson, M. J. (2004). A phylogeny of legumes (Leguminosae) based on analysis of the plastid matK gene resolves many well-supported subclades within the family. *American Journal of Botany*, 91, 1846-1862.

[119] Taberlet, P., Gielly, L., Pautou, G., & Bouvet, J. (1991). Universal primers for amplification of three non-coding regions of chloroplast DNA. *Plant Molecular Biology*, 17, 1105-1109.

[120] Wolfe, A. D., Elisens, W. J., Watson, L. E., & Depamphilis, C. W. (1997). Using restriction-site variation of PCR-amplified cpDNA genes for phylogenetic analysis of Tribe Cheloneae (Scrophulariaceae). *American Journal of Botany*, 84, 555-564.

[121] Sehgal, D., & Raina, S. N. (2008). DNA markers and germplasm resource diagnostics: new perspectives in crop improvement and conservation strategies. *In: Arya ID, Arya S. (ed.) Utilization of biotechnology in plant sciences*, Microsoft Printech (I) Pvt. Ltd., Dehradun, 39-54.

[122] Varshney, R. K., Graner, A., & Sorrells, M. E. (2005). Genic microsatellite markers in plants: features and applications. *Trends Biotechnology*, 23, 48-55.

[123] Varshney, R. K., Thudi, M., Aggarwal, R., & Börner, A. (2007). Genic molecular markers in plants: development and applications. *In: Varshney RK, Tuberosa R (ed.) Genomics assisted crop improvement: genomics approaches and platforms*, Springer, Dordrecht, 1, 13-29.

[124] Powell, W., Machray, G. C., & Provan, J. (1996). Polymorphism revealed by simple sequence repeats. *Trends Plant Sciences*, 1, 215-222.

[125] Russell, J. R., Fuller, J. D., Macaulay, M., Hatz, B. G., Jahoor, A., Powell, W., & Waugh, R. (1997). Direct comparison of levels of genetic variation among barley accessions detected by RFLPs, AFLPs, SSRs and RAPDs. *Theoretical and Applied Genetics*, 95, 714-722.

[126] Bohn, M., Utz, H. F., & Melchinger, A. E. (1999). Genetic similarities among winter wheat cultivars determined on the basis of RFLPs, AFLPs, and SSRs and their use for predicting progeny variance. *Crop Sciences*, 39, 228-237.

[127] Parida, S. K., Kalia, S. K., Sunita, K., Dalal, V., Hemaprabha, G., Selvi, A., Pandit, A., Singh, A., Gaikwad, K., Sharma, T. R., Srivastava, P. S., Singh, N. K., & Mohapatra, T. (2009). Informative genomic microsatellite markers for efficient genotyping applications in sugarcane. *Theoretical and Applied Genetics*, 118, 327-338.

[128] Shimamoto, Y., Hasegawa, A., Abe, J., Ohara, M., & Mikami, T. (1992). Glycine soja germplasm in Japan: isozymes and chloroplast DNA variation. *Soybean Genet Newsletter*, 19, 73-77.

[129] Xu, D. H., Abe, J., Gai, J. Y., & Shimamoto, Y. (2002). Diversity of chloroplast DNA SSRs in wild and cultivated soybeans: evidence for multiple origins of cultivated soybean. *Theoretical and Applied Genetics*, 105, 645-653.

[130] Xu, D. H., Abe, J., Kanazawa, A., Gai, Y., & Shimamoto, Y. (2001). Identification of sequence variations by PCR-RFLP and its application to the evaluation of cpDNA diversity in wild and cultivated soybeans. *Theoretical and Applied Genetics*, 102, 683-688.

[131] Hirata, T., Abe, J., & Shimamoto, Y. (1996). RFLPs of chloroplast and motochondrial genomes in summer and autumn maturing cultivar groups of soybean in Kyushu district of Japan. *Soybean Genetic Newsletter*, 23, 107-111.

[132] Shimamoto, A., Fukushi, H., Abe, J., Kanazawa, A., Gai, J., Gao, Z., & Xu, D. (1998). RFLPs of chloroplast and mitochondrial DNA in wild soybean, Glycine soja, growing in China. *Genetic Resources and Crop Evolution*, 45, 433-439.

[133] Chen, J., & Hebert, P. D. N. (1999). Directed termination of the polymerase chain re-action: kinetics and application in mutation detection. *Genome*, 42, 72-79.

[134] Orita, M., Suzuki, T. S., & Hayashi, K. (1989). Rapid and sensitive detection of point mutation and DNA polymorphisms using the polymerase chain reaction. *Genomics*, 5, 874-879.

[135] Sarkar, G., Yoon, H., & Sommer, S. S. (1995). Dideoxy fingerprinting (ddF): a rapid and efficient screen for the presence of mutations. *Genomics*, 13, 441-443.

[136] Manen, J. F., & Natall, A. (1995). Comparison of the evolution of ribulose- 1, 5-bi-sphosphate carboxylase (rbcL) and atpB-rbcL non-coding spacer sequences in a re-cent plant group, the tribe Rubieae (Rubiaceae). *Journal of Molecular Evolution*, 41, 920-927.

[137] Jordan, W. C., Courtney, M. W., & Neigel, J. E. (1996). Low levels of intraspecific ge-netic variation at a rapidly evolving chloroplast DNA locus in North American duck-weeds (Lemnaceae). *American Journal of Botany*, 83, 430-439.

[138] Sang, T., Crawford, D. J., & Stuessy, T. F. (1997). Chloroplast DNA phylogeny, reticu-late evolution, and biogeography of Paeonia (Paeoniaceae). *American Journal of Bot-any*, 84, 1120-1136.

[139] Small, R. L., Ryburn, J. A., Cronn, R. C., Seelanan, T., & Wendel, J. F. (1998). The tor-toise and the hare: choosing between noncoding plastome and nuclear Adh sequen-ces for phylogeny reconstruction in a recently diverged plant group. *American Journal of Botany*, 85, 1301-1315.

[140] Mc Dade, L. A., & Moody, M. L. (1999). Phylogenetic relationship among Acantha-ceae: evidence from noncoding trnL-trnFF chloroplast DNA sequences. *American Journal of Botany*, 86, 70-80.

[141] Molvray, M., Kores, P. J., & Chase, M. W. (1999). Phylogenetic relationships within Korthalsella (Viscaceaek) based on nuclear ITS and plastid trnL-F sequence data. *American Journal of Botany*, 86, 249-260.

[142] Demesure, B., Comps, B., & Petit, R. J. (1996). Chloroplast DNA phylogeography of the common beach (Fagus sylvatica L.) in Europe. *Evolution*, 50, 2515-2520.

[143] Wolfe, A. D., Elisens, W. J., Watson, L. E., & Depamphilis, C. W. (1997). Using restric-tion-site variation of PCR-amplified cpDNA genes for phylogenetic analysis of Tribe Cheloneae (Scrophulariaceae). *American Journal of Botany*, 84, 555-564.

[144] Asmussen, C. B., & Liston, A. (1998). Chloroplast DNA characters, phylogeny, and classification of Lathyrus (Fabaceae). *American Journal of Botany*, 85, 387-40.

[145] Cipriani, G., Testolin, R., & Gardner, R. (1998). Restriction-site variation of PCR-am-plified chloroplast DNA regions and its implication for the evolution of Actinidia. *Theoretical and Applied Genetics*, 96, 389-396.

[146] Friesen, N., Pollner, S., Bachmann, K., & Blattner, F. R. (1999). RAPDs and noncoding chloroplast DNA reveal a single origin of the cultivated Allium fistulosum from A. altaicum (Alliaceae). *American Journal of Botany*, 86, 554-562.

[147] Close, P. S., Shoemaker, R. C., & Keim, P. (1989). Distribution of restriction site polymorphism within the chloroplast genome of the genus Glycine, subgenus Soja. *Theoretical and Applied Genetics*, 77, 768-776.

[148] Kanazawa, A., Tozuka, A., & Shimamoto, Y. (1998). Sequence variation of chloroplast DNA that involves EcoRI and ClaI restriction site polymorphisms in soybean. *Genes Genetic Systems*, 73, 111-119.

[149] Downie, S. R., Olmstead, R. G., Zurawski, G., Soltis, D. E., Soltis, P. S., Watson, C., & Palmer, J. D. (1991). Six independents losses of the chloroplast DNA rpl2 intron in dicotyledons: Molecular and phylogenetic implications. *Evolutions*, 45, 1245-1259.

[150] Spielmann, A., & Stutz, E. (1983). Nucleotide sequence of soybean chloroplast DNA regions which contain the psbA and trnH genes and cover the ends of the large single-copy region and one end of the inverted repeats. *Nucleic Acids Research*, 11, 7157-7167.

[151] Sugiura, M., & Shimada, H. (1991). Fine- structural features of the chloroplast genome- comparison of the sequenced chloroplast genomes. *Nucleic Acids Research*, 19, 983-995.

[152] Taberlet, P., Coissac, E., Popanon, F., Gielly, L., Miquel, C., Valentini, A., Vermat, T., Corthier, G., Brochmann, C., & Willerslev, E. (2006). Power and limitations of the chloroplast trnL (UAA) intron for plant DNA barcoding. *Nucleic Acids Research*, 35(3).

[153] Palmer, J. D. (1991). Plastid chromosomes: Structure and Evolution. *In: Bogorag L., Vasil IK (ed.) Cell culture and somatic cell genetic of plants*, New York, Academic Press, 7, 5-53.

Chapter 8

An Overview of Genetic Transformation of Soybean

Hyeyoung Lee, So-Yon Park and Zhanyuan J. Zhang

Additional information is available at the end of the chapter

1. Introduction

Soybean (*Glycine max* (L.) Merrill) is a model legume crop, widely grown in the world for human consumption or animal fodder. Moreover, soybeans have gained worldwide research interest in many public laboratories and industrial sectors. Soybean seeds contain protein, oil, carbohydrates, dietary fibers, vitamins, and minerals. For the last few decades the majority of research laboratories have been investigating genetic traits to improve the yield of protein or oil in soybean seeds through genetic engineering, thereby achieving improved quantity and quality of soybean seeds. Until now, most of the transformation experiments have implemented a single functional gene not multiple genes. Those agronomically and economically important traits affect the enhancement of grain quantity and quality [1]. However, the majority of agronomic and genetic traits such as complex metabolic, biological, and pharmaceutical pathways are polygenetic traits and are produced in a complex pathway. Therefore, those traits are encoded and regulated by a number of genes. In an attempt to study and manipulate those pathways, the transfer of multigene or large inserts into plants have been developed by multigene engineering technology and have also been involved in metabolic engineering. Several examples of multigene or large insert transfers have been reported such as the application of carotenogenic genes in rice, canola, and maize [2-4], and of polyunsaturated fatty acid and vitamin E genes in soybean and *Arabidopsis* [5-7]. Therefore, reliable systems for transforming large DNA fragments into plants make it feasible to introduce a natural gene cluster or a series of previously unlinked foreign genes into a single locus.

Over the last two decades, the transfer of DNA into plant cells has been achieved by using several methods. In soybeans, the most frequently employed plant genetic engineering methods are *Agrobacterium*-mediated transformation and particle bombardment. Both systems have successfully been used in genetic transformation of soybean. Since the initial reports of fertile transgenic soybean production [8-9], various efforts have been made to improve the transformation efficiency and to produce transgenic soybean. Particularly, the preferred and reproducible transformation is the use of the cotyledonary node as a plant material, which is based on *Agrobacterium*-mediated gene transfer [10-12]. Nevertheless, new methods have been developed for more efficient soybean transformation. There still remain, however, many challenges for genotype- and tissue- specific independent transformation of soybean. This review provides an overview of historical efforts in developing and advancing soybean regeneration and transformation systems. In addition, recent advances and challenges in soybean transformation are discussed.

2. Different approaches for soybean transformation

In soybean transformation, two major methods are now widely utilized: *Agrobacterium-mediated* transformation of different explant tissues and particle bombardment. The *Agrobacterium*-mediated method, as a simple protocol, does not require any specific or expensive equipment. Moreover, this method usually produces single or low copy numbers of insertions with relatively rare rearrangement [13]. On the other hand, bombardment technique directly introduces desired genes into the target plant cell with small tungsten or gold particles [9]. The success of this approach critically depends upon the ability of the target tissue to proliferate as well as proper pre-cultures to make a target plant.

2.1. Cotyledonary-node-based transformation

The routine regeneration system was first reported by using the mature cotyledonary-node [14]. The multiple adventitious buds and shoots from explant tissues were proliferated and regenerated on culture media containing cytokinin by organogenesis. The transgenic soybean plants have been successfully and reproducibly produced using mature or immature cotyledon explants via *Agrobacterium*-mediated transformation. Hinchee et al. [8] for the first time reported the production of fertile transgenic soybean plants using mature cotyledonary-node by *Agrobacterium*-mediated transformation, but transformation efficiency was very low. The system employed the neomycin phosphotransferase II (*NPT* II) gene as a selectable marker and combined kanamycin as a selective agent. However, this selection was addressed with a problem of regeneration of non-transgenic or chimeric shoots at the shoot formation stage. Moreover, the system was highly genotype-dependent. To overcome the high genotype-dependency and high chimerism problems by the *NPT* II selection and develop a new selection system for soybean transformation, Zhang et al. [10] developed the selection system employing herbicide bialaphos resistance (*bar*) gene as a selectable marker coupled with glufosinate as a selective agent. This system enabled to transform many soybean genotypes with stable transgene inheritance, albeit transformation efficiency remained to be improved. Meanwhile, to solve the escape problem caused by kanamycin selection, Clemete et al. [15] deployed the herbicide glyphosate as a selective agent, leading to high stringent selection and good transgene inheritance. It was discovered later that addition of various thiol compounds in the co-cultivation medium significantly increased the transformation efficiency [11, 16-17]. These thiol compounds, as antioxidants, reduce the oxidative burst that caused tissue browning or necrosis and also promote organogenesis and shoot growth from buds [18].

Recently, an alternative cotyledonary explant derived from mature soybean seed for *Agro-bacterium* transformation has been reported by Paz et al. [19]. The term half-seed explants were used as an experiment material and fertile transgenic plants were attained.

In fact, several laboratories have contributed to enhanced soybean transformation using a cotyledonary-node explant. To overcome the low transfer of *Agrobacterium* into plant cell, the infection media were first amended with the phenolic compound, 4'-Hydroxy-3',5'-dime-thoxyacetophenone (acetosyringone), to induce expression of the virulence (*Vir*) genes [20-21]. To increase the infection sites, Trick and Finer [22] evaluated cotyledonary node transformation efficiency using a developed sonication assisted *Agrobacterium*-mediated transformation (SAAT) protocol. Although this treatment was not able to obtain fertile transgenic plants, the increase of *Agrobacterium* transfer was shown. Olhoft et al. [16-17] discovered that thiol compounds enhanced *Agrobacterium* infection in soybean. At the same time, however, these compounds caused counter-selection effect when glufosinate was used as a selective agent under previously published selection conditions. To solve this problem, Olhoft et al. [11] developed Hygromycin phosphotransferase (*HPT* II) selection system using hygromycin B as a selective agent. This has led to a substantial increase in transformation frequency. Transformation efficiency with thiol compounds was increased 5-fold by using refined glufosinate selection [12].

Since the transformation process by use of kanamycin or hygromycin B as selection agent has been proven to be genotype-dependent, the most widely used selection system has been the combination of *bar* gene with the herbicide phosphinothricin (glufosinate) [10, 12]. In this selection system, the concentration of agent glufosinate greatly affects the transformation frequency [12], so the appropriate selection schemes can be varied among genotypes, seed vigor and other *in vitro* culture conditions.

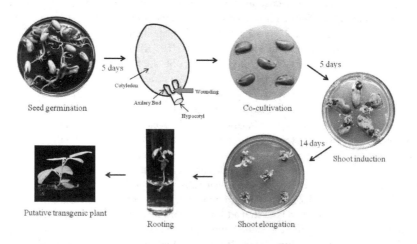

Figure 1. Scheme for genetic transformation of soybean (*Glycine max* (L.) Merrill) cotyledonary nodes.

2.2. Immature embryos-based transformation

The regeneration using immature embryos via somatic embryogenesis was first reported by Christianson et al. [23]. The immature embryos excised from soybean pods were suspended on semi-solid media or liquid media containing high concentration of auxin, 2,4-Dichlorophe-noxyacetic acid (2,4-D), and the whole plantlets were recovered [24-25]. After immature embryos were developed as an alternative plant material, transgenic plants were first obtained from this explant tissue via particle bombardment [26]. This system has been exclusively used to produce transgenic soybean such as glyphosate tolerant, hygromycin resistance, and *Bacillus thuringiensis* (BT) transgenic soybean [27-29]. As the formation of proliferative embryogenic tissue depends on genotype, the use of immature embryos for transformation has been limited to few genotypes cultivars including "Jack" and "Williams 82."

The use of particle bombardment with immature embryos tends to be highly variable, and multiple copies of the introduced DNAs are commons. Moreover, this problem has compounded with aged embryogenic suspension cultures from which a high percentage of regenerated plants lost their fertility [29]. In spite of this limitation, the embryogenic cultures have several advantages, one of which is its relatively high transformation efficiency and less chimeric plants recovered.

2.3. Embryogenic shoot tips-based transformation

The embryonic shoot tip explant is another source of explant which has been used for soybean transformation. McCabe et al., [30] first reported the stable transformation using meristemic cell, shoot apex, by particle acceleration. The shoot derived from these meristems via organogenesis has been produced to form multiple shoots prior to mature plants. However, all of the primary transgenic plants were chimeric. Martinell et al., [31] described the successful method using meristemic shoot tip from germinated seedling by *Agrobacterium*-mediated transformation. This system has provided rapid and efficient soybean transformation. Liu et al. [32] also reported the regeneration system using embryonic shoot tips by shoot organogenesis. The explants have been shown the high regeneration and the transformation efficiencies using *Agrobacterium*-mediated with up to 15.8%.

2.4. Immature cotyledonary-nodes

The regeneration capacity of immature cotyledonary-node was found by Parrott et al [33]. Based on this regeneration system, first transgenic soybean plants have been developed by *Agrobacterium tumefaciens* [34]. This system was tested using two different *Agrobacterium* strains, LBA4404 and EHA101 and deploying kanamycin selection. The system utilized auxin 1-Naphthaleneacetic acid (NAA) for plant regeneration. Although these systems allowed development of transgenic plants from the explants, no fertile transgenic plants were recovered. Recently, Ko et al [35] described the efficient transformation system using immature cotyledonary-nodes by *Agrobacterium*-mediated transformation, but transformation efficiency was still very low.

2.5. Hypocotyl based transformation

Another type of explant tissue, hypocotyl, was also investigated with 13 different soybean genotypes. Most of the genotypes initiated shoots from this type of explant [36]. This method was reported to be genotype-independent regeneration protocol via organogenesis and utilized the acropetal end of a hypocotyl section from a 7-day old seedling. Despite inducing adventitious shoots from the explant, most recovered shoot did not matured in the soil. Wang et al [37] reported successful production of fertile transgenic plants using hypocotyl-based *Agrobacterium*-mediated transformation. To improve the transformation system, two different chemicals, cytokinin hormone 6-Benzylaminopurine (BAP) and silver nitrate, were added to the shoot formation media. In spite of the term "hypocotyl" used in the above transformation system, the true tissues responsible for regeneration are actually the preexisting meristem tissues located at the nodal area of the cotyledon, essentially the same source of tissue as cotyledonary-nodes [17] except that cotyledons were removed [45, 46].

2.6. Leaf tissue-based transformation

The reproducible regeneration methods for whole plants from primary leaf tissue or epicotyls were first reported by Wright et al [38]. The multiple shoots from those explants were continually initiated and proliferated with cytokinin BAP hormone. Rajasckaren et al [39] described regeneration of several varieties of soybean by embryogenesis from epicotyls and primary leaf tissues, thereby inducing fertile plants from those explants. Kan et al [40] first tested transformation efficiency using epicotyls and leaf tissues by *Agrobacterium tumefaciens*. To find out proper transformation condition for those explants, they investigated different *Agrobacterium* strains, EHA101 and LBA4404, but also different treatments on inoculation stage, sucrose and mannose.

3. *Agrobacterium*-mediated transformation of soybean

3.1. *Agrobacterium*-mediated transformation mechanism

Agrobacterium is a unique organism to generate transgenic plants and in natural conditions [41]. It allows introduction of a single stranded copy of the bacterial transferred DNA (T-DNA) into a host cell and integration of the genomic DNA of interest, resulting in genetic manipulation of the host. Since the development of disarmed tumour-inducing (Ti) plasmid [42-43], *Agrobacterium* has been used to transform various major crops for genetic modification [44-46].

Agrobacterium recognizes wounded host plant cells which produce penolic compounds such as acetosyringone as inducers of *vir* gene expression [47], and attach to the plant cells to export the T-DNA after virulence (Vir) protein activation. Acetosyringone is now routinely used for improving transformation efficiency. After *vir* gene activation, a single stranded T-DNA copy (T-strand) is transferred into the plant by type IV secretion system (T4SS) which is related to VirB complex [48]. The VirB complex is composed of at least 12 proteins (VirB1-11 and VirD4) which form a multisubunit envelope-spanning structure [49]. Various

Agrobacterium proteins, such as VirD2-T-DNA, VirE2, VirE3, VirF, and VirD5, pass though VirB complex to transfer into plant cells [50-51]. VirE2 and VirD2 interact with cytosolic T-DNA in the plant cells and form a complex which is later imported into the nucleus when it is bound to VIP1 plant protein [52-55]. Recently, Gelvin et al., hypothesized that T-complex (T-DNA, VirE2, VirD2 and VIP1) is imported into the nucleus through actin cytoskeleton and thus myosin may be involved in *Agrobacterium*-mediated transformation [56]. However, the specific mechanism of T-DNA movement through myosin is still unknown.

The T-complex is imported into the nucleus by the phosphorylation of VirE2 Interacting Protein 1 (VIP1), induced by mitogen-activated protein kinase (MAPK), such as MPK3 [55]. After T-complex is imported into the host nucleus, VirE2 and VIP1 need to be degraded before T-DNA integration by a subunit of the SCF (SKP-CUL1-F-box protein) ubiquitin E3 ligase complex. Not only *Agrobacterium* protein VirF but also protein VBF can mark VIP1 protein for the degradation. Furthermore, binding of VIP1-binding F-box (VBF) to T-complex can induce the degradation of VIP1 and VirE2 by the 26S proteosome, and at the end T-strand is integrated into plant genomic DNA and expressed in the host plants [57-58].

3.2. History of *Agrobacterium*-mediated soybean transformation research

Among various transformation technologies, *Agrobacterium*-mediated transformation method has shown to be effective for the production of transgenic soybeans because of straightforward methodology, familiarity to researchers, minimal equipment cost, and reliable insertion of a single transgene or a low copy number [13]. Till now, a number of reports have been published related to the optimum condition to achieve a high yield of soybean transformation; such as *Agrobacterium* inoculation conditions, regeneration media components, etc. For *Agrobacterium*-mediated transformation methods, the susceptibility of soybean to *Agrobacterium* and various *Agrobacterium* strains have been tested to improve the transformation efficiency (Table 1). Also, *Agrobacterium* strains and growth conditions which affect the soybean transformation efficiency have been published [8, 59-62]. After Pederson et al., [46] and Owens et al., [59] showed the susceptibility of certain soybean genotypes against tumor induction, *Agrobacterium* biology study has been advanced to enhance transformation efficiency. In addition to *Agrobacterium* biology study, chemical contents for inoculation have been studied such as varying acetosyringone and syringaldehyde concentrations [63]. For high inoculation efficiency, Mauro et al., [64] tested various *Agrobacterium* biotypes (nopaline, agropine and octopine) to identify the most effective *Agrobacterium* biotype for soybean transformation.

After Hinchee et al., [8] developed *Agrobacterium*-mediated soybean transformation methods, many *Agrobacterium* strains have been tested and employed, such as EHA101, EHA105, LBA4404 and AGL1. Parrott et al., [33] showed that EHA101 was highly potent to transform immature soybean cotyledons, especially PI283332, and had higher recovery of transformed plants over LBA4404. Dang and Wei [65] tested transformation efficiency using embryonic tips instead of cotyledonary explants and somatic embryos, and when embryogenic tips were infected for 20 hours, hypervirulent strain KYRT1 showed increased efficiency over EHA105 and LBA4404.

Recently, *A. tumefaciens* KAT23 (AT96-6) which has an ability to efficiently transfer the T-DNA into soybean, was isolated from peach root. After 20 stains were confirmed by common bean and soybean transformation, Yukawa et al. [66] tested their potential availability as legume super virulent *A. tumefaciens* in various soybean cultivars (Peking, Suzuyutaka, Fayette, Enrei, Mikawashima, WaseMidori, Jack, Leculus, Morocco, Serena, Kentucky Wonder and Minidoka). Without modifying vectors or *vir* function, they showed that KAT23 (AT96-6) has a high potential to function as a common strain to increase soybean transformation efficiency. Therefore, this study identified a novel soybean super virulent *A. tumefaciens* strain which transferred not only the T-DNA of the Ti-plasmid but also introduced T-DNA of the binary vector efficiently. These results indicate that KAT23 (AT96-6) has the ability to transform soybeans at high efficiency.

There has been a significant improvement in soybean transformation over the past two decades. However, the efficiency of soybean transformation is not great enough for practical needs and shows high variation. Thus, considering the potential application of soybean transformation, the importance of *Agrobacterium* can't be over-emphasized.

Strain of *A. tumefaciens*	Soybean genotype	Selection		Reference
		Marker	Agent	
A208	Peking, Maple Prest	*npt* II	kanamycin	(8)
AGL1	Bert	*bar*	phosphinothricin	(67)
EHA101	Williams 82	bar	glufosinate	(12)
EHA101	Williams, Williams 79, Peking, Thorne	*bar*	glufosinate or bialaphos	(68)
EHA101	Thorne, Williams, Williams 79, Williams 82	*bar*	glufosinate	(19)
EHA105	AC Colibri	*npt* II	kanamycin	(69)
EHA105	Hefeng 25, Dongnong 42, Heinong 37, Jilin 39, Jiyu 58	*hpt*	hygromycin	(70)
EHA105	A3237	*bar*	glufosinate	(10)
LBA4404	Jungery	*bar*	phosphinothricin	(71)
LBA4404, EHA105	Bert	*hpt*	hygromycin	(11)

Table 1. Summary of cotyledonary-node transformation system.

4. New directions of soybean genetic engineering, skills and vectors

To date, the *Agrobacterium*- and biolistic-mediated transformation methods remain the very successful methods in soybean transformation, whereas other available transformation technologies have not been practical in soybean, which include electroporation-mediated transformation [72], PEG/liposome-mediated transformation [73], silicon carbide-mediated transformation [74], microinjection [75] and chloroplast-mediated transformation [76]. Of these two, *Agrobacterium*-mediated transformation has become more adapted in public laboratories worldwide. On the other hand, there are unintended insertions such as unwanted antibiotic markers and promoters, which can be inserted during transformation. This problem has raised potential biosafety issues related to environmental concerns and human health risks. To overcome these potential risks, methods of developing marker free transgenic plants have been developed, such as cotransformation [77], transposon-mediated transformation [78] and site-specific recombination [79].

Among the various methods, co-transformation system is one of the most commonly used methods to produce marker free transgenic plants. In co-transformation systems, a marker gene and genes of interest are placed on separate DNA molecules and introduced into plant genomes. Then, the non-selectable genes segregate from the marker gene in the progeny generations. Most strains of *A. tumefaciens* have the ability to contain more than one T-DNA, and crown gall tumors were often co-transformed with multiple T-DNAs [42]. As a result, there are two possibilities; Multiple T-DNAs were delivered into plant cells either from a mixture of strains ('mixture methods') or from a single strain ('single-strain methods'). Depicker et al. [80] described that a single strain method was higher in efficiency than a mixture method. For a single-strain method of co-transformation, Kamori et al., [77] tested co-transformation method to develop a suitable superbinary vector system. Using the unique plasmids which carried two T-DNA segments marker free rice and tobacco were produced and evaluated. LBA4404, a derivative of an octopine strain, were used for these co-transformation methods and they hypothesized that LBA4404 may be an important factor contributing to the high frequency of unlinked loci.

To improve plant genetic traits, many soybean research labs have developed tools for soybean functional genomics, such as several libraries containing large inserts of bacterial artificial chromosome (BAC) and plant transformation competent binary plasmids clone (BIBAC) (81). In functional genomic research, bacterial artificial chromosome (BAC) is a single copy artificial chromosome vector and is based on the *E. coli* fertility (F-factor) plasmid. They are not only stable in host cell, but also are used for large scale gene cloning and discovery [82]. However, some BAC libraries that are desirable for functional genomics are often not amenable for transformation directly into plants because of their large subclones. Therefore, binary bacterial artificial chromosome (BIBAC) libraries have been developed for *Agrobacterium*-mediated plant transformation and gene functional complementation. The BIBAC library is based on BAC vector and has both an F-factor plasmid for replication origin of *E.coli* and an Ri plasmid for replication origin of *Agrobacterium rhizogenes*. The vector also has a *sacB* gene as a positive selection for *E. coli* and a selectable marker gene for plant. Since BIBAC vectors were reported, these vectors have been used for plant transformation in some model plant species including tobacco, canola, tomato, and rice [83-86]. Although the transformation efficiency was very low, the BIBAC vectors have been successfully employed to transfer large inserts into those crops as a single locus via *Agrobacterium*-mediated transformation. The introduced T-DNA was stably maintained and inherited through several generations and no gene silencing was observed [85]. However, soybean transformation using a BIBAC vector has not been achieved to date. Moreover, plant transformation with DNA fragments below 20 kb is routine whereas the stable plant transformation with DNA fragments larger than 50 kb is challenging.

5. Zinc finger nucleases (ZFNs) and transcription activator-like effectors (TALEs)

Although many methods have been developed, soybean is considered a recalcitrant plant to transform compared to *Arabidopsis* and rice. Since full genome sequencing data has been rapidly updated in soybean, soybean transformation technology is becoming an essential approach for genomic research. For the phenotypic analysis of genes, knock-out or gene-silencing plants are used to study gene function. However in soybean, making bulk knock-out mutants through conventional mutagenesis approaches is not immediately feasible because of low transformation efficiency. Thus, development of innovative gene targeting methods is necessary to make knock-out plants in soybean.

Zinc finger nucleases (ZFNs) and meganucleases cut specific DNA target sequences *in vivo* and thus are powerful tools for genome modification. In particular zinc finger domain, which predominantly recognize nucleotide triplets, have been widely used in this research. Importantly, ZFNs modification has been reported in soybean (*Glycine max*), maize (*Zea mays*), tobacco (*Nicotiana tabacum*) and *Arabidopsis* [87-90]. Unfortunately, Ramirez et al. [91] found a major disadvantage of ZFNs; they observed that the modular assembly method of engineering zinc-finger arrays has a higher failure rate than previously reported.

To overcome the ZFNs's weakness, in late 2009, a novel DNA binding domain was identified which was a member of the large transcription activator-like (TAL) effector family [92-93]. Transcription activator-like effectors (TALEs) are produced by plant pathogens in the genus *Xanthomonas* as virulence factors and TAL effector–mediated gene induction leads to plant developmental changes [94-95]. The type III secretion system is used by *Xanthomonas* to introduce virulence factors into plant cells [96]. Once inside the plant cell, transcription activator-like (TAL) effectors (TALEs) enter the nucleus, bind effector-specific DNA sequences, and transcriptionally activate gene expression [97-98]. For genomic engineering, two methods of TALEs were developed: TALE nucleases (or TALENs) and TALE transcription factors (or TALE-TFs). Both TALENs and TALE-TFs contain as many as 30 tandem repeats of a 33- to 35-amino-acid-sequence motif (Figure 2). The amino acids in positions 12 and 13 in each 33- to 35-amino-acid-sequence motif have the repeat variable di-residue (RVD). Using this specific ability, two pair (left and right TALENs) of repeats with different RVDs are designed by PCR and bound in the target DNA sequence [92-93]. Fok1 combined with TALE nucleases (or TALENs) make double-strand breaks (DSBs) at specific locations in the genome. These DSBs are repaired by homologous recombination (HR) or non-homologous end-joining (NHEJ) pathways. During DSBs repair, errors in genome via insertion, deletion, or chromosomal rearrangement could be induced by HR and NHEJ (Figure 2). Unlike TALENs, TALE-TFs require only a single TALE construct for activity induction when combined with VP64 activator (derived from the herpex simplex virus activation domain). VP64 binds with RNA polymerase and causes transcriptional activation of the gene of interest [99]. For making transgenic plants, the TALEs technique can be combined with the *Agrobacterium*-mediated transformation method and it is assumed that gene targeting knock-out will be applied in soybean research.

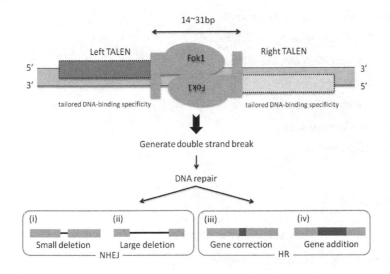

Figure 2. Summary of Transcription activator-like effectors (TALEs) nuclease.

6. Abbreviation

Acetosyringone: 4'-Hydroxy-3',5'-dimethoxyacetophenone

BAC: bacterial artificial chromosome

bar: bialaphos resistance

BAP: 6-Benzylaminopurine

2,4-D: 2,4-Dichlorophenoxyacetic acid

BIBAC: binary bacterial artificial chromosome

BT: *Bacillus thuringiensis*

DSBs: double-strand breaks

HPT II: Hygromycin phosphotransferase

HR: homologous recombination

NAA: Naphthaleneacetic acid

NHEJ: non-homologous end-joining

NPT II: neomycin phosphotransferase II

RVD: repeat variable di-residue

Ti plasmid: tumour-inducing (Ti) plasmid

T4SS: type IV secretion system

TALEs: transcription activator-like effectors

VBF: VIP1-binding F-box

VIP1: VirE2 Interacting Protein 1

Vir protein: virulence protein

ZFNs: Zinc finger nucleases

Author details

Hyeyoung Lee, So-Yon Park and Zhanyuan J. Zhang*

*Address all correspondence to: zhangzh@missouri.edu

Plant Transformation Core Facility, University of Missouri, Columbia, USA

References

[1] Mazur, B., Krebbers, E., & Tingey, S. (1999). Gene Discovery and Product Development for Grain Quality Traits. *Science*, 285(5426), 372-375.

[2] Ye, X., et al. (2000). Engineering the provitamin A (beta-carotene) biosynthetic pathway into (carotenoid-free) rice endosperm. (Translated from eng). *Science*, 287(5451), 303-305, (in eng).

[3] Paine, J. A., et al. (2005). Improving the nutritional value of Golden Rice through increased pro-vitamin A content. (Translated from eng). *Nat Biotechnol*, 23(4), 482-487, (in eng).

[4] Fujisawa, M., et al. (2009). Pathway engineering of Brassica napus seeds using multiple key enzyme genes involved in ketocarotenoid formation. (Translated from eng). *J Exp Bot*, 60(4), 1319-1332, (in eng).

[5] Kinney, A. J., et al. (2004). *COMPANY EIDPDNA WO/2004/071467*.

[6] Qi, B., et al. (2004). Production of very long chain polyunsaturated omega-3 and omega-6 fatty acids in plants. (Translated from eng). *Nat Biotechnol*, 22(6), 739-745, (in eng).

[7] Karunanandaa, B., et al. (2005). Metabolically engineered oilseed crops with enhanced seed tocopherol. (Translated from eng). *Metab Eng*, 7(5-6), 384-400, (in eng).

[8] Hinchee, M. A. W., et al. (1988). Production of Transgenic Soybean Plants Using Agrobacterium-Mediated DNA Transfer. *Nat Biotech*, 6(8), 915-922.

[9] Christou, P., Mc Cabe, D. E., & Swain, W. F. (1988). Stable Transformation of Soybean Callus by DNA-Coated Gold Particles. *Plant Physiology*, 87(3), 671-674.

[10] Zhang, Z., Xing, A., Staswick, P., & Clemente, T. E. (1999). The use of glufosinate as a selective agent in Agrobacterium-mediated transformation of soybean. *Plant Cell, Tissue and Organ Culture* , 56(1), 37-46.

[11] Olhoft, P., Flagel, L., Donovan, C., & Somers, D. (2003). Efficient soybean transformation using hygromycin B selection in the cotyledonary-node method. *Planta*, 216(5), 723-735.

[12] Zeng, P., Vadnais, D. A., Zhang, Z., & Polacco, J. C. (2004). Refined glufosinate selection in Agrobacterium-mediated transformation of soybean [Glycine max (L.) Merrill]. *Plant Cell Reports*, 22(7), 478-482.

[13] Hansen, G., & Wright, M. S. (1999). Recent advances in the transformation of plants. *Trends in Plant Science*, 4(6), 226-231.

[14] Wright, M. S., Koehler, S. M., Hinchee, M. A., & Carnes, M. G. (1986). Plant regeneration by organogenesis in Glycine max. *Plant Cell Reports*, 5(2), 150-154.

[15] Clemente, T. E., et al. (2000). Progeny Analysis of Glyphosate Selected Transgenic Soybeans Derived from-Mediated Transformation. *Crop Sci.*, 40(3), 797-803.

[16] Olhoft, P. O., Lin, K. L., Galbraith, J. G., Nielsen, N. N., & Somers, D. S. (2001). The role of thiol compounds in increasing Agrobacterium-mediated transformation of soybean cotyledonary-node cells. *Plant Cell Reports*, 20(8), 731-737.

[17] Olhoft, P. O., & Somers, D. S. (2001). L-Cysteine increases Agrobacterium-mediated T-DNA delivery into soybean cotyledonary-node cells. *Plant Cell Reports*, 20(8), 706-711.

[18] Dan, Y. (2008). Biological functions of antioxidants in plant transformation. *In Vitro Cellular & Developmental Biology- Plant*, 44(3), 149-161.

[19] Paz, M., Martinez, J., Kalvig, A., Fonger, T., & Wang, K. (2006). Improved cotyledonary node method using an alternative explant derived from mature seed for effi-

cient <i>Agrobacterium</i>-mediated soybean transformation. *Plant Cell Reports*, 25(3), 206-213.

[20] Stachel, S. E., Messens, E., Van Montagu, M., & Zambryski, P. (1985). Identification of the signal molecules produced by wounded plant cells that activate T-DNA transfer in Agrobacterium tumefaciens. *Nature*, 318(6047), 624-629.

[21] Owens, L. D., & Smigocki, A. C. (1988). Transformation of Soybean Cells Using Mixed Strains of Agrobacterium tumefaciens and Phenolic Compounds. (Translated from eng). *Plant Physiol*, 88(3), 570-573, (in eng).

[22] Trick, H. N., & Finer, J. J. (1997). SAAT: sonication-assisted Agrobacterium-mediated transformation. *Transgenic Research*, 6(5), 329-336.

[23] Christianson, M. L., Warnick, D. A., & Carlson, P. S. (1983). A Morphogenetically Competent Soybean Suspension Culture. *Science*, 222(4624), 632-634.

[24] Ranch, J., Oglesby, L., Zielinski, A., & Horsch, R. (1985). Plant regeneration from embryo-derived tissue cultures of soybeans. *In Vitro Cellular & Developmental Biology - Plant*, 21(11), 653-658.

[25] Finer, J. J., & Nagasawa, A. (1988). Development of an embryogenic suspension culture of soybean (Glycine max Merrill.). *Plant Cell, Tissue and Organ Culture* , 15(2), 125-136.

[26] Finer, J., & Mc Mullen, M. (1991). Transformation of soybean via particle bombardment of embryogenic suspension culture tissue. *In Vitro Cellular & Developmental Biology - Plant*, 27(4), 175-182.

[27] Padgette, S. R., et al. (2009). Development, Identification, and Characterization of a Glyphosate-Tolerant Soybean Line. *Crop Sci.*, 35(5), 1451-1461.

[28] Hadi, M. Z., Mc Mullen, M. D., & Finer, J. J. (1996). Transformation of 12 different plasmids into soybean via particle bombardment. *Plant Cell Reports*, 15(7), 500-505.

[29] Stewart, C. N., Jr., et al. (1996). Genetic Transformation, Recovery, and Characterization of Fertile Soybean Transgenic for a Synthetic Bacillus thuringiensis cryIAc Gene. *Plant Physiology*, 112(1), 121-129.

[30] Mc Cabe, D. E., Swain, W. F., Martinell, B. J., & Christou, P. (1988). Stable Transformation of Soybean (Glycine Max) by Particle Acceleration. *Nat Biotech*, 6(8), 923-926.

[31] Martinell, B. J., Horeb, M., Julson, L. S., Mills, L., Emler, C. A., Huang, Y., Mc Cabe, D. E., & Williams, E. J. (2002). *Patent No. US 6384301*.

[32] Liu, H.-K., Yang, C., & Wei, Z.-M. (2004). Efficient Agrobacterium tumefaciens-mediated transformation of soybeans using an embryonic tip regeneration system. *Planta*, 219(6), 1042-1049.

[33] Parrott, W., Williams, E., Hildebrand, D., & Collins, G. (1989). Effect of genotype on somatic embryogenesis from immature cotyledons of soybean. *Plant Cell, Tissue and Organ Culture*, 16(1), 15-21.

[34] Parrott, W. A., Hoffman, L. M., Hildebrand, D. F., Williams, E. G., & Collins, G. B. (1989). Recovery of primary transformants of soybean. *Plant Cell Reports*, 7(8), 615-617.

[35] Ko-S, T., Lee, S., Krasnyanski, S., & Korban, S. S. (2003). Two critical factors are required for efficient transformation of multiple soybean cultivars: <i>Agrobacterium</i> strain and orientation of immature cotyledonary explant. *TAG Theoretical and Applied Genetics* , 107(3), 439-447.

[36] Dan, Y., & Reichert, N. (1998). Organogenic regeneration of soybean from hypocotyl explants. *In Vitro Cellular & Developmental Biology- Plant*, 34(1), 14-21.

[37] Wang, G., & Xu, Y. (2008). Hypocotyl-based Agrobacterium-mediated transformation of soybean (Glycine max) and application for RNA interference. *Plant Cell Reports*, 27(7), 1177-1184.

[38] Wright, M. S., Williams, M. H., Pierson, P. E., & Carnes, M. G. (1987). Initiation and propagation of Glycine max L. Merr.: Plants from tissue-cultured epicotyls. *Plant Cell, Tissue and Organ Culture*, 8(1), 83-90.

[39] Rajasekaran, K., & Pellow, J. (1997). Somatic embryogenesis from cultured epicotyls and primary leaves of soybean [<i>Glycine max</i> (L.) Merrill]. *In Vitro Cellular & Developmental Biology- Plant*, 33(2), 88-91.

[40] Khan, R. (2006). AG SP.

[41] De Cleene, M., & De Ley, J. (1976). The host range of crown gall. *The Botanical Review*, 42(4), 389-466.

[42] Hooykaas, P. J. J., & Schilperoort, R. A. (1992). Agrobacterium and plant genetic engineering. *Plant Molecular Biology*, 19(1), 15-38.

[43] Newell, C. (2000). Plant transformation technology. *Molecular Biotechnology*, 16(1), 53-65.

[44] Ishida, Y., et al. (1996). High efficiency transformation of maize (Zea mays L.) mediated by Agrobacterium tumefaciens. *Nat Biotech*, 14(6), 745-750.

[45] Hiei, Y., Komari, T., & Kubo, T. (1997). Transformation of rice mediated by Agrobacterium tumefaciens. *Plant Molecular Biology*, 35(1), 205-218.

[46] Pedersen, H. C., Christiansen, J., & Wyndaele, R. (1983). Induction and in vitro culture of soybean crown gall tumors. *Plant Cell Reports*, 2(4), 201-204.

[47] Stachel, S. E., & Nester, E. W. (1986). The genetic and transcriptional organization of the vir region of the A6 Ti plasmid of Agrobacterium tumefaciens. (Translated from eng). *EMBO J*, 5(7), 1445-1454, (in eng).

[48] Cascales, E., & Christie, P. J. (2003). The versatile bacterial type IV secretion systems. (Translated from eng). *Nat Rev Microbiol*, 1(2), 137-149, (in eng).

[49] Christie, P. J., Atmakuri, K., Krishnamoorthy, V., Jakubowski, S., & Cascales, E. (2005). Biogenesis, architecture, and function of bacterial type IV secretion systems. (Translated from eng). *Annu Rev Microbiol*, 59, 451-485, (in eng).

[50] Vergunst, A. C., et al. (2000). VirB/D4 -dependent protein translocation from Agrobacterium into plant cells. (Translated from eng). *Science*, 290(5493), 979-982, (in eng).

[51] Vergunst, A. C., et al. (2005). Positive charge is an important feature of the C-terminal transport signal of the VirB/D4 -translocated proteins of Agrobacterium. (Translated from eng). *Proc Natl Acad Sci U S A*, 102(3), 832-837, (in eng).

[52] Gelvin, S. B. (2003). Agrobacterium-mediated plant transformation: the biology behind the "gene-jockeying" tool. (Translated from eng). *Microbiol Mol Biol Rev*, 67(1), 16-37, table of contents (in eng).

[53] Christie, P. J., Ward, J. E., Winans, S. C., & Nester, E. W. (1988). The Agrobacterium tumefaciens virE2 gene product is a single-stranded-DNA-binding protein that associates with T-DNA. (Translated from eng). *J Bacteriol*, 170(6), 2659-2667, (in eng).

[54] Citovsky, V., Zupan, J., Warnick, D., & Zambryski, P. (1992). Nuclear localization of Agrobacterium VirE2 protein in plant cells. (Translated from eng). *Science*, 256(5065), 1802-1805, (in eng).

[55] Djamei, A., Pitzschke, A., Nakagami, H., Rajh, I., & Hirt, H. (2007). Trojan horse strategy in Agrobacterium transformation: abusing MAPK defense signaling. (Translated from eng) . *Science*, 318(5849), 453-456, (in eng).

[56] Gelvin, S. B. (2012). Traversing the cell: Agrobacterium T-DNAs journey to the host genome. (Translated from English). *Frontiers in Plant Science*, 3, (in English).

[57] Zaltsman, A., Krichevsky, A., Loyter, A., & Citovsky, V. (2010). Agrobacterium induces expression of a host F-box protein required for tumorigenicity. (Translated from eng). *Cell Host Microbe*, 7(3), 197-209, (in eng).

[58] Tzfira, T., Vaidya, M., & Citovsky, V. (2004). Involvement of targeted proteolysis in plant genetic transformation by Agrobacterium. (Translated from eng). *Nature*, 431(7004), 87-92, (in eng).

[59] Owens, L. D., & Cress, D. E. (1985). Genotypic variability of soybean response to agrobacterium strains harboring the ti or ri plasmids. (Translated from eng). *Plant Physiol*, 77(1), 87-94, (in eng).

[60] Delzer, B. W., Somers, D. A., & Orf, J. H. . Agrobacterium tumefaciens Susceptibility and Plant Regeneration of 10 Soybean Genotypes in Maturity Groups 00 to II. *Crop Sci.*, 30(2), 320-322.

[61] Olhoft, P. M., & Somers, D. A. (2007). *Soybean Transgenic Crops VI. Biotechnology in Agriculture and Forestry, eds Pua E-C & Davey MR*, Springer, Berlin, Heidelberg, 61, 3-27.

[62] Somers, D. A., Samac, D. A., & Olhoft, P. M. (2003). Recent Advances in Legume Transformation. *Plant Physiology*, 131(3), 892-899.

[63] Smigocki, A. C., & Owens, L. D. (1988). Cytokinin gene fused with a strong promoter enhances shoot organogenesis and zeatin levels in transformed plant cells. (Translated from eng). *Proc Natl Acad Sci U S A*, 85(14), 5131-5135, (in eng).

[64] Mauro, A. O., Pfeiffer, T. W., & Collins, G. B. (2009). Inheritance of Soybean Susceptibility to Agrobacterium tumefaciens and Its Relationship to Transformation. *Crop Sci.*, 35(4), 1152-1156.

[65] Dang, W., & Wei, Z-m. (2007). An optimized Agrobacterium-mediated transformation for soybean for expression of binary insect resistance genes. *Plant Science*, 173(4), 381-389.

[66] Yukawa, K., Kaku, H., Tanaka, H., Koga-Ban, Y., & Fukuda, M. (2007). Characterization and host range determination of soybean super virulent Agrobacterium tumefaciens KAT23. *Bioscience, biotechnology, and biochemistry* , 71(7), 1676-1682.

[67] Olhoft, P. O., & Somers, D. S. (2001). L-Cysteine increasesAgrobacterium-mediated T-DNA delivery into soybean cotyledonary-node cells. *Plant Cell Reports*, 20(8), 706-711.

[68] Paz, M., et al. (2004). Assessment of conditions affecting Agrobacterium-mediated soybean transformation using the cotyledonary node explant. *Euphytica*, 136(2), 167-179.

[69] Donaldson, P. A., & Simmonds, D. H. (2000). Susceptibility to Agrobacterium tumefaciens and cotyledonary node transformation in short-season soybean. *Plant Cell Reports*, 19(5), 478-484.

[70] Liu, S.-J., Wei, Z.-M., & Huang, J.-Q. (2008). The effect of co-cultivation and selection parameters on Agrobacterium-mediated transformation of Chinese soybean varieties. *Plant Cell Reports*, 27(3), 489-498.

[71] Xue, R.-G., Xie, H.-F., & Zhang, B. (2006). A multi-needle-assisted transformation of soybean cotyledonary node cells. *Biotechnology Letters*, 28(19), 1551-1557.

[72] Salmenkallio-Marttila, M., et al. (1995). Transgenic barley (Hordeum vulgare L.) by electroporation of protoplasts. *Plant Cell Reports*, 15(3), 301-304.

[73] Davey, M. R., Anthony, P., Power, J. B., & Lowe, K. C. (2005). Plant protoplasts: status and biotechnological perspectives. *Biotechnology Advances*, 23(2), 131-171.

[74] Kaeppler, H. F., Somers, D. A., Rines, H. W., & Cockburn, A. F. (1992). Silicon carbide fiber-mediated stable transformation of plant cells. *TAG Theoretical and Applied Genetics*, 84(5), 560-566.

[75] Crossway, A., et al. (1986). Integration of foreign DNA following microinjection of tobacco mesophyll protoplasts. *Molecular and General Genetics MGG*, 202(2), 179-185.

[76] Boynton, J., et al. (1988). Chloroplast transformation in Chlamydomonas with high velocity microprojectiles. *Science*, 240(4858), 1534-1538.

[77] Komari, T., Hiei, Y., Saito, Y., Murai, N., & Kumashiro, T. (1996). Vectors carrying two separate T-DNAs for co-transformation of higher plants mediated by Agrobacterium tumefaciens and segregation of transformants free from selection markers. *The Plant Journal*, 10(1), 165-174.

[78] Yan, H., & Rommens, C. M. (2007). Transposition-Based Plant Transformation. *Plant Physiology*, 143(2), 570-578.

[79] Bai, X., Wang, Q., & Chu, C. (2008). Excision of a selective marker in transgenic rice using a novel Cre/ <i>loxP</i> system controlled by a floral specific promoter. *Transgenic Research*, 17(6), 1035-1043.

[80] Depicker, A., Herman, L., Jacobs, A., Schell, J., & Montagu, M. (1985). Frequencies of simultaneous transformation with different T-DNAs and their relevance to the <i>Agrobacterium</i>/plant cell interaction. *Molecular and General Genetics MGG*, 201(3), 477-484.

[81] Wu, C., et al. (2004). A BAC- and BIBAC-based physical map of the soybean genome. (Translated from eng). *Genome Res*, 14(2), 319-326, (in eng).

[82] Zhang-B, H., & Wu, C. (2001). BAC as tools for genome sequencing. *Plant Physiology and Biochemistry*, 39(3-4), 195-209.

[83] Hamilton, C. M., Frary, A., Lewis, C., & Tanksley, S. D. (1996). Stable transfer of intact high molecular weight DNA into plant chromosomes. (Translated from eng). *Proc Natl Acad Sci U S A*, 93(18), 9975-9979, (in eng).

[84] Cui, Y., Bi, Y. M., Brugiere, N., Arnoldo, M., & Rothstein, S. J. (2000). The S locus glycoprotein and the S receptor kinase are sufficient for self-pollen rejection in Brassica. (Translated from eng). *Proc Natl Acad Sci U S A*, 97(7), 3713-3717, (in eng).

[85] Frary, A., & Hamilton, C. M. (2001). Efficiency and stability of high molecular weight DNA transformation: an analysis in tomato. (Translated from eng). *Transgenic Res*, 10(2), 121-132, (in eng).

[86] He, R. F., et al. (2006). Development of transformation system of rice based on binary bacterial artificial chromosome (BIBAC) vector. (Translated from eng). *Yi Chuan Xue Bao*, 33(3), 269-276, (in eng).

[87] Shukla, V. K., et al. (2009). Precise genome modification in the crop species Zea mays using zinc-finger nucleases. *Nature*, 459(7245), 437-441.

[88] Townsend, J. A., et al. (2009). High-frequency modification of plant genes using engineered zinc-finger nucleases. *Nature*, 459(7245), 442-445.

[89] Curtin, S. J., et al. (2011). Targeted Mutagenesis of Duplicated Genes in Soybean with Zinc-Finger Nucleases. *Plant Physiology*, 156(2), 466-473.

[90] Zhang, F., et al. (2010). High frequency targeted mutagenesis in Arabidopsis thaliana using zinc finger nucleases. *Proceedings of the National Academy of Sciences*, 107(26), 12028-12033.

[91] Ramirez, C. L., et al. (2008). Unexpected failure rates for modular assembly of engineered zinc fingers. *Nat Meth*, 5(5), 374-375.

[92] Boch, J., et al. (2009). Breaking the Code of DNA Binding Specificity of TAL-Type III Effectors. *Science*, 326(5959), 1509-1512.

[93] Moscou, M. J., & Bogdanove, A. J. (2009). A Simple Cipher Governs DNA Recognition by TAL Effectors. *Science*, 326(5959), 1501.

[94] Bogdanove, A. J., Schornack, S., & Lahaye, T. (2010). TAL effectors: finding plant genes for disease and defense. *Current Opinion in Plant Biology*, 13(4), 394-401.

[95] Kay, S., & Bonas, U. (2009). How Xanthomonas type III effectors manipulate the host plant. *Current Opinion in Microbiology*, 12(1), 37-43.

[96] Boch, J., & Bonas, U. (2010). Xanthomonas AvrBs3 Family-Type III Effectors: Discovery and Function. *Annual Review of Phytopathology*, 48(1), 419-436.

[97] Gürlebeck, D., Szurek, B., & Bonas, U. (2005). Dimerization of the bacterial effector protein AvrBs3 in the plant cell cytoplasm prior to nuclear import. *The Plant Journal*, 42(2), 175-187.

[98] Römer, P., Recht, S., & Lahaye, T. (2009). A single plant resistance gene promoter engineered to recognize multiple TAL effectors from disparate pathogens. *Proceedings of the National Academy of Sciences*, 106(48), 20526-20531.

[99] Sanjana, N. E., et al. (2012). A transcription activator-like effector toolbox for genome engineering. *Nat. Protocols*, 7(1), 171-192.

Chapter 9

Comparative Studies Involving Transgenic and Non-Transgenic Soybean: What is Going On?

Marco Aurélio Zezzi Arruda,
Ricardo Antunes Azevedo,
Herbert de Sousa Barbosa,
Lidiane Raquel Verola Mataveli,
Silvana Ruella Oliveira,
Sandra Cristina Capaldi Arruda and
Priscila Lupino Gratão

Additional information is available at the end of the chapter

1. Introduction

There is no doubt that soybean [*Glycine max* (L.) Merrill] has a global importance with widespread applicability (food, biodiesel, secondary metabolites, among others) and economic value of its products in the global market [1] (*ca.* US$ 38.9 billion is the estimated crop value in 2010 [2]). In terms of crop production, USA and Brazil occupy the first and second position in the world, with *ca.* 83 and 70 million ton, respectively [3]. These facts may explain the significant interest by biotechnological industries and research institutes in enhancing some characteristics of this crop such as nutrient quality, resistance to pests, and other subjects.

One of the most effective processes for attaining this task is the genetic modification, and clearly the one conferring resistance to glyphosate (the number one selling herbicide worldwide since the 80's [4]) is considered the main soybean genetic modification. This process involves the insertion of a gene (cp4 EPSPS: 5-enolpyruvylshikimate-3 phosphate synthase), via biobalistic (acceleration of metallic particles recovered with genetic material), which is responsible to the production of cp4-EPSPS enzyme from *Agrobacterium sp*. Such enzyme confers tolerance to glyphosate (N-fosfometil glycine), once that this substance inhibits the action of 5-enolpyruvylshiquimate synthase – EPSPS enzyme, which is involved in the biosynthesis of aromatic amino acids. Its inhibition provokes delay in the development of plants, amino acids unbalance, and death of plants [5]. As a consequence it is easy to rationalize that plants genetically modified can normally develop in the presence of this herbicide, being the excellent result obtained in terms of soybean crop production also attributed to the transgenic cultures used. Exemplifying, *ca.* 49 million hectares (60% of the cultivated world area) are occupied nowadays by transgenic soybean culture [6].

In this way, our hypothesis is that the genetic modification itself is contributing for changing a variety of characteristics of this organism, producing alterations, in a cascade manner, to the metabolism. As a result of these modifications, the genetically modified soybean is apparently searching a new equilibrium as a living organism in nature.

In order to evaluate this hypothesis, our research group has been, for the last eight years, carrying out some comparative studies taking into account alterations in proteins, metalloproteins, metals and enzymes. Since the content of proteins in soybean seeds is high (*ca.* 40%) [7], hundreds of proteins are expected to be found after a separation process, making our proposal a hard task. Then, the utilization of the most up-to-date analytical techniques presenting high resolution in terms of enzymes, proteomics and metallomics approaches is almost imperative.

In this chapter, the concept of plant stress is mainly one related to oxidative stress, followed by a variety of examples regarding soybean. Additionally, basic concepts of proteomics and metallomics will be described, followed by a compilation of the results from all strategies and techniques that we have been adopting along the period devoted to the study of transgenic soybean, which were utilized for corroborating our previous hypothesis. Other examples in the literature are also presented in order to support our data. Then, techniques based on bidimensional chromatographic and non-chromatographic protein separations (*i.e.* 2D-HPLC, 2D PAGE), image analysis for protein expression evaluations (*i.e.* 2D DIGE), inorganic mass spectrometry for identification/quantification of metals (*i.e.* HR-SF-ICP-MS, ICP-MS, LA-ICP-MS), organic mass spectrometry for characterization of proteins (*i.e.* MALDI-QTOF-MS, ESI-LC-MS-MS), and hyphenated techniques for improving the quality on protein information (*i.e.* 2D-HPLC-ICP-MS) will be also emphasized. In the end of this chapter, a section of future Trends is provided, putting in evidence, in our point of view, some other strategies to be adopted for an in-depth investigation of this transgenic crop.

Finally, it is important to stress that the main goal of this chapter, and also of our studies, is only to present those results found within a series of projects developed by our research group concerning transgenic soybean. Despite the awareness of a public disagreement about the cultivation and commercialization of transgenic soybean, this chapter does not have the intention neither to defend genetic modification nor to make any criticisms to it.

2. Plant stress

In this section we have decided to focus our attention on the antioxidant responses trigged by some key biotic and abiotic stresses that have more significant information available, based on recent publications.

When oxidative stress is taken into account, it is interesting to mention firstly, the role of molecular oxygen (O_2) in our environment. Due to the presence of oxygen and its reactions, both positive and negative aspects inherent to the process can occur, which is called oxidative stress.

Among all planets in our solar system, Earth is the only one that contains O_2, and the only one able to support aerobic life as the way that we understand its meaning. According to [8], the concentration of 21% (v/v) O_2 on Earth's atmosphere is derived from the photosynthetic activities of cyanobacterias and plants. The reference [9] commented that by an estimate, the total amount of O_2 in the Earth is about 410×10^3 Erda moles and from this value, 38.4×10^3 Erda moles is in the water form. When the aerobic life is concerned, these authors commented that this specific style of life is responsible for the major portion of O_2 turnover: photosynthesis is the main input of O_2, and respiration the main output.

Oxygen is relatively non reactive, but in some situations (as normal metabolic activity or when under environmental disturbance), it is able to switch to an excited state, producing free radicals and similar forms [9-10]. Then, it is clear in this scenario that adaptation processes to environmental changes are crucial for plant growth and survival. In view of its importance, it is interesting to remember the processes which lead to the reduction of molecular O_2. According to [11] such processes occur following four steps and generate several O_2 species. The first one requires an extra energy but the subsequent steps are exothermic, occurring spontaneously. The reaction products (H_2O_2; $O_2^{\bullet}$; $HO_2^{\bullet}$; OH^-) can act in different ways in the cellular environment.

Hydrogen peroxide is a relatively long-lived molecule and can diffuse from its site of production [12]. Beside this, its toxicity has long been known. The $O_2^{\bullet}$ radical half-life is short (2-4 µs), but it is highly reactive and can form hydroperoxides and can oxidize histidine, methionine and tryptophan. When this radical is in the cellular environment, it causes lipid peroxidation as a consequence of oxidative deterioration of membrane polyunsaturated lipids. So, the hydrogen peroxide is not only toxic to cells, but in an extracellular medium it may react with transition metals, such as iron and copper, generating hydroxyls, which can cause cell damage. Beside this, when the levels of lipid peroxidation are higher (normally lipid peroxidation values are estimated by the concentration of malondialdehyde in samples) it suggests indirectly the establishment of a condition of oxidative stress. The hydroxyl radical (OH^-) has a very strong potential and half-life of less than 1µs, and as a consequence, it has very high affinity for biological molecules [13]. What is particularly interesting about these species is that all of them can be generated by molecular oxygen reduction and they may play roles as toxic molecules or they can be excellent candidates for events/studies involving plant cell signaling [8, 13-16].

In these terms, the production of reactive oxygen species (ROS) is generally described as harmful due to their potential to cause irreversible damage to photosynthetic components in plants. However, despite this potential in causing harmful oxidation, modulation of ROS-antioxidant interaction plays a role in many stresses, as well as other responses to the environment. Additionally, this system can be considered as a powerful signaling process to molecules involved in the control of plant growth and development as well as priming acclimatory responses to stress stimuli [17-18]. In these terms, oxidative stress can be described as a central factor in abiotic and biotic stress that occurs due to imbalances in any cell compartment between the production of ROS and antioxidant defense [16, 19].

As indicated in [18] it is possible to verify that the pathways of ROS signaling are made by homeostatic regulation which can be achieved by the antioxidant redox buffering, making possible the determination of lifetime and the specificity of the ROS signal. It is interesting to emphasize that plants which demonstrate low activities for catalase (CAT) and cytosolic ascorbate peroxidase (APX), two key enzymes involved in the breakdown of H_2O_2, show less severe stress symptoms when compared to the ones where one of these enzymes is missing [20].

Talking about antioxidant defense systems, it can be attested that, in plants, the first line of defense against oxidative stress is the avoidance of ROS production [17] and once formed, ROS must be detoxified in order to either avoid or minimize eventual damages. In this way, the detoxification mechanisms can be considered as a second line of defense against the detrimental effects of ROS [21]. Beside this, some antioxidant enzymes can be considered as a second defense line against oxidative stress, since they act either as a catalyzer in ROS reaction or are involved in directing ROS processing [22]. The repair of oxidatively damaged proteins can be considered as the third line of defense against ROS [23].

According to [24] ROS species are commonly generated under stress conditions and due to its strong oxidative capacity, it acts on all types of biomolecules. In terms of the interactive effects of these species, it is possible to say that it can react with each other and with other molecules. For example, $O_2^{\bullet-}$ may react with lipids peroxides or nitric oxide, leading to the formation of peroxynitrite, which is less reactive than peroxides. In the same context, [13] pointed out that plants may favor the formation of one or other reactive species by preferentially scavenging peroxide (H_2O_2) with antioxidants or, in contrast, accumulating peroxide by the activation of superoxide dismutase (SOD).

The oxidative response in plants can be exacerbated by stressful conditions [16]. At the molecular level, the extent and nature of this response can differ among species and even among those closely related varieties of the same species. For example, 24 differentially expressed genes in soybean leaves were observed after glyphosate treatment when comparing tolerant and non tolerant soybean lines [25]. Therefore, oxidative responses are not only linked to the genetic expression. The reference [26] shows that some biochemical parameters (such as total soluble amino acid content and CAT activity in soybean roots) were also altered as a response to differential glyphosate application. The increase in the enzyme activities indicates ROS generation and a subsequent antioxidant response. Alterations in the antioxidative system of suspension-cultured soybean cells were observed [27], which were induced by oxidative stress using a peroxidizing herbicide (oxyfluorfen). Ascorbate and glutathione (non-enzymatic cellular antioxidants) showed different responses and the activities of some enzymes involved in cellular defense were also altered. For instance, peroxidase and catalase increased by 40 – 70% while glutathione S-transferase (GST) exhibited a 6-fold increase under oxyfluorfen stress.

Stress-induced ROS accumulation is counteracted either by enzymatic oxidant systems that include a variety of scavengers, such as superoxide dismutase (SOD), ascorbate peroxidase (APX), glutathione peroxidase (GPX), glutathione S-transferase (GST), and catalase (CAT), or by non enzymatic low molecular weight metabolites, such as carotenoids and flavonoids [16, 28-29]. As an example related to the influence of the enzymatic machinery under a stress situation, the reference [30] pointed out that as a response to stress, plants may increase the activities of some enzymes such as glutathione S-transferase (GST), involved in the detoxification of xenobiotics. These authors also investigated in detail the mechanisms of interaction between the GST enzyme and its substrates, indicating that the information might help in the engineering of new GSTs with improved detoxification efficiency [30].

In the context so far referred, and particularly with soybean, which is the main focus of this chapter, molecular and biochemical studies have explored several aspects related to the manipulation of metabolic pathways towards adaptation responses which can help to mitigate oxidative stress.

If ROS scavenging pathways in plants are the main focus, the involvement of at least 3 cycles have to be considered: a) the water-water cycle in chloroplasts, including SOD; b) the ascorbate-glutathione cycle in chloroplasts, cytosol, mitochondria, appoplast and peroxisomes; and c) GPX and CAT in the peroxisomes [31]. The equilibrium between the production and the scavenging of ROS may be altered by biotic and abiotic stress factors such as UV radiation, temperature, air pollution, pathogen attack, heavy metals, nutrient deficiency, and herbicides, among others [32].

The clear understanding of the mechanisms by which some endogenous or exogenous agents can lead to plant toxicity and how plants answer to this specific situation, is essential. Besides this, understanding how toxicity occurs, what kind of alterations occur in plant structure and metabolism among other situations are important steps for genetic breeding programs, when searching for new varieties susceptible or tolerant to stress factors and even for bioremediation/phytoremediation programs [32].

Although there has been a rapid progress in recent years in the field of plant stress studies, there is a consensus among researches that there are still many uncertainties in understanding how effectively ROS affects the stress response of plants [32-33].

A short list of examples that will emphasize the detoxification mechanism involved in plant stress defense and a diversity of enzymes that can be involved in the dismutation of ROS is then presented. For example, soybean has been shown to be highly sensitive to ozone (O_3) and the oxidation of some proteins may cause alterations in the activities of enzymes across nitrogen and sulfur nutrient assimilation pathways linked to stress responses [34]. The chronic exposure to high O_3 may lead to increased expression or oxidation of proteins, including APX, GSTs [34] and decrease the activities of monodehydroascorbate reductase (MDHAR) and glutathione reductase (GR) [35], indicating a fundamental role of these enzymes in stress response when soybean is subjected to O_3. Furthermore, soybean submitted to chronic high O_3 concentration and then exposed to an acute O_3 stress provided evidence that there was an immediate transcriptional reprogramming that allowed for maintained or increased ascorbate (AA) content in plants grown at high O_3 [36]. In another study using two tropical soybean varieties (PK 472 and Bragg) exhibiting differential sensitivity to O_3, reference [37] showed that the CAT activity decreased whereas peroxidase increased in both varieties upon exposure to O_3, but reflecting the greatest sensitivity of PK 472 in relation to the high magnitude of the reductions in the levels of antioxidants, metabolites and nutrients. Besides this, the damage O_3 effects produced were found to be more prominent during the reproductive than the vegetative growth stage.

Although soybean plants have shown a positive and significant correlation between activity of antioxidant enzymes and the osmolyte proline (Pro) content to water deficit stress [38], the metabolic reasons associated with the differential sensitivity of soybean cultivars to water deficit stress are not well understood [39]. According to the authors, water deficit stress increased antioxidant enzyme activities of SOD, CAT and GPX more at mild than at high water deficit stress [39]. In addition, soybean plants have shown protective mechanisms associated to proline concentration and GR, APX, and CAT activities under salt stress [40].

The increase in soybean productivity has been also accounted to the development and widespread use of improved cultivars with increased resistance to stressful conditions. A promising technique for agricultural improvement in arid and semiarid areas is the use of a pretreatment of soybean dry seeds with a low dose of gamma rays (20 Gy) before planting, enhancing drought tolerance and minimizing the yield losses caused by a water deficit condition [41].Overall, application of a low dose of gamma irradiation (20 Gy) increased the activities of phosphoenol pyruvate carboxylase and ribulose-1,5-bisphosphate carboxylase/oxygenase (RUBISCO) under drought stress, avoiding the destructive effects of water deficits on chloroplasts [41]. Furthermore, the manipulation of Pro can affect the (h)GSH, amino acids concentrations and APX activity, contributing to the detoxification of ROS in soybean subjected to simultaneous drought and heat stresses [42].

The regulation of thiol metabolism has become important for optimizing crop yield and quality of soybean [43]. The sulfur assimilatory pathway in soybean metabolism can be metabolized into molecules that protect plants against oxidative stress. The genetic manipulation of the cytosolic isoform of O-acetylserine sulfhydrylase (OASS), an enzyme involved in the sulfur assimilatory pathway, resulted in high levels of thiols and increased tolerance of plants to metal toxicity [44].

It is also important that information concerning changes in antioxidant capacity in immature seeds harvested at different reproductive stages [45], exhibited decreases in free radical scavenging activity and total antioxidant capacity with the advancement of maturity. This occurred concomitant with increased concentration of tocopherol and isoflavone isomers. Therefore, it is important to take into consideration that not only organs or tissues may present distinct responses to stress, but the plant stage of development is also important.

Reference [46] reported that different metals may act and induce different levels of copper-zinc superoxide dismutase (Cu-Zn/SOD) expression in soybean plants exposed to Cd and Pb. Also, Cd caused the induction of Cu-Zn/SOD mRNA accumulation for all Cd concentrations and Pb-treated roots showed induction of these isoenzymes only at medium metal concentrations.

It is also important to bring into the scene of stressful condition for plants, the soil type used. In a recent work soybean exposed to Cd and Ba [47] showed that the activity of antioxidant enzymes can change depending on the soil type, time-length of exposure and metal concentration [47]. For instance, GR and SOD activities in the leaves of soybean plants grown in an Oxisol soil contaminated with Cd decreased over time, whilst remaining high in an Entisol soil. The changes of enzyme activities were mainly dependent on buffering capacity of the soils with the Entisol exhibiting a lower capacity, with the plants suffering higher oxidative stress than those plants grown in a clay soil such as presented by an Oxisol soil [47].

Moreover, it is also important to investigate the effect of stressful conditions in soybean productivity taking into consideration more than one environmental contaminant or stress factor in the same agricultural region. In this context, the combination of Cd and acid rain pollution damaged the cell membrane, decreased the activities of POX and CAT, showing a higher potential threat to soybean seed germination than the single separate effect of each contaminant [48]. In another study, a correlation between the rate of ROS generation and antioxidant enzyme activities was established under hypoxia and high CO_2 concentration [49]. The CAT activity in soybean plants increased during the first hours of hypoxia whereas peroxidase activity started to play a more key role in cell defense only after a longer exposure to hypoxia. In this study, the processes of ROS accumulation and antioxidant enzymes were induced by the higher CO_2 content, indicating that CO_2 can switch on plant adaptation to hypoxic stress [49].

Another interesting study involving the combination of distinct stressor agents was carried out by combining Al and Cd with both leading to synergistic effects on plant growth and antioxidant responses in two soybean cultivars with different Al tolerance levels [50]. According to the authors, the Al treatments and low pH value (4.0) caused reduction in chlorophyll content and net photosynthetic rate, leading to growth reduction. The increased SOD and peroxidase activities were detected in the plants submitted to both metals, especially in the Al-sensitive cv. Zhechun 2, which also exhibited significantly higher Al and Cd contents than the Al tolerant cv. Liao-1. Moreover, Cd supplementation increased Al content in the plants exposed to Al+Cd stress [50]. Such an observation confirms another key aspect that should receive attention which is how the elements of the soil interact and can define an uptake profile by the plant root system possibly resulting in an induction of stress condition. Such studies are also of the upmost importance when considering the use of phytoremediation as a technique to recover a contaminated soil.

Similarly, studies about the interactions between plant roots and beneficial metal-tolerant microorganisms are gaining importance and may be an important approach to be considered in studies about plant adaptation and alleviation to a variety of environmental stresses [51]. For example, soybean plants inoculated with arbuscular mycorrhizal (AM) fungi showed reduced MDA content and increased APX activity to the oxidative stress generated by paraquat (PQ) [52]. In another study, activities of SOD and peroxidase were increased in the shoots of soybean plants with mycorrhizal (M) fungi grown under

NaCl salinity [53]. Once again, a more integrated view is needed and deserves attention. These two studies commented above indicate the importance of mycorrhizal fungi regulation as a general strategy to protect plants from stress. If soil type is added to this equation, a much more complex situation is created and such an integrated study reflects the reality of many agronomic situations. This also raises the question over the use of hydroponic systems to study oxidative stress in plants, particularly when induced by non-essential elements, since it is not necessarily the real field condition. Yet, is not our intention to say that such studies under hydroponic conditions are not important. On the contrary, they also have advantages. However, a more dynamic or integrated type of study should be considered in our point of view.

Curiously, grafting, which is a well-known agronomic technique largely used in agriculture, has not been used much in studies of stress in plants. The grafting technique has a tremendous potential to add further important understanding about stress signaling, assimilation, transport and accumulation of metals, opening a new perspective to study these grafted plants at the biochemical and molecular levels. Unfortunately, very few examples are available in the literature focusing on the investigation of plant stress responses. An example of such a study is the one carried out by [54] who showed that Cd seed concentration can be influenced by the difference in translocation of Cd from soil to the seed and in Cd accumulation capacity of roots among soybean cultivars by the use of grafting.

Nowadays, the development of plant manipulation techniques, for example the production and use of transgenic plants, has contributed to studies involving plant antioxidant responses induced either by exogenous or endogenous factors (such as herbicide, metals, pollution). Studies involving the mechanisms leading to stress-tolerant plants are important and needed, since they can aid understanding and create new possibilities for the use of these kinds of plants. The knowledge provided by the "omics studies" such as proteomics, metabolomics, metallomics and genomics, added to enzymatic evaluation, can provide information that can decisively help in answering many questions related to oxidative stress and ROS control [32].

Taking into account the importance of "omics" platforms, as well as their use for corroborating our initial hypothesis, the following sections will focus on these important strategies. They will be divided into proteomics and metallomics, with brief descriptions of each one, as well as some discussions and examples regarding transgenic cultures, but always concentrating the focus on soybean.

3. Proteomics

Proteomics can be defined as the large-scale study of proteins, including not only their identifications and quantifications, but also the determination of their localizations, modifications, interactions, activities, and functions [55]. This information is extremely important to evaluate interactions between different proteins, or between proteins and other molecules, and may reveal the functional role of proteins [56]. In this sense, proteomics is an important part of plant science, providing essential tools for understanding the functions of many plant-specific biological processes at the molecular level [57]. Currently, plant proteomic studies are focused on understanding the impact of different conditions of plant physiology such as the characterization of plant defense under biotic and abiotic stress [58,59], the characterization of subcellular, cellular or plant organ proteomes [60-61], the characterization of genetic modifications [62-63], as well as others.

The insertion of exogen DNA fragments into the DNA of the target organism, to confer some enhanced characteristics to the latter, describes the process termed genetic modification [64]. Focusing on plants, improved productiveness, enhanced tolerance to herbicides, synthesis of new substances and others can have consequences related to genetic modification [64]. The natural responses to this process are known to change the protein map of an organism [65]. In this sense, comparative proteomics become the strategy of choice, being useful for establishing qualitative and quantitative differences between genetically and non-genetically modified organisms [66]. In this way, studies of protein changes are frequently carried out through polyacrylamide gels by evaluation of their images, providing relevant information for comparative proteomic studies [67,68] as well as using appropriate mass spectrometric techniques for evaluating the identity of the studied proteins [69-70].

For proteomic studies, gel electrophoresis separations are the most used platform, due to their high resolution, allowing either high efficiency protein separation or the identification of potential protein spots with differences in concentration or expression in the gels evaluated [67,68]. The gel electrophoresis technique can be applied in proteomic studies as follow: (1) one-dimensional gel electrophoresis (SDS-PAGE) [71], (2) two-dimensional gel electrophoresis (2-D PAGE) followed by manual image analysis [66] and (3) two-dimensional difference gel electrophoresis (2-D DIGE) followed by automatic image analysis [72]. The application of these techniques in comparative studies involving transgenic soybeans has been little explored, where variation of different proteomic profiles in soybean genotypes [73], abiotic environmental stress [74], osmotic stress [75] and improvement of protein quality in transgenic soybean plants [76] are examples found. In this way, the use of these separation techniques in combination with mass spectrometry were applied in our research group to comparative proteomic studies in transgenic and non-transgenic soybean seeds, and it will be discussed below.

4. One-dimensional gel electrophoresis (SDS-PAGE)

The separation of the proteins using SDS-PAGE technique is based on their molecular mass, covering a broad separation range [67]. In our research group, this technique was used with mass spectrometry (ESI-QTOF MS/MS) for the identification of the enzyme cp4 EPSPS, in order to prove that the soybean in question was genetically modified. For this task, the protein band corresponding to a mass of 47 kDa was cut, the proteins reduced, alkylated, and subjected to two enzymatic digestion protocols: trypsin and chymotrypsin.

As a result, the enzyme cp4 EPSPS was identified using the SDS-PAGE technique and using either trypsin or chymotrypsin as a cleavage enzyme. However, trypsin showed the best results in terms of score and coverage (as a percentage). Moreover, the enzyme was identified in the database containing sequences from the soil bacterium *Agrobacterium* sp, the origin of the gene used in genetic modification. This method proved to be simple and very efficient, without needing sample prefractionation using chromatographic columns [77].

5. Two-dimensional gel electrophoresis: 2-D PAGE

In 2-D PAGE technique, the proteins are first separated on the basis of their net charges by isoelectric focusing (IEF) and then separated on the basis of their molecular mass by polyacrylamide gel electrophoresis in the presence of sodium dodecyl sulfate (SDS) [67].

The combination of 2-D PAGE and mass spectrometry is one excellent strategy to obtain proteomic maps [78]. Furthermore, taking into account that we were finding some differential proteins when investigating transgenic and non-transgenic soybean seeds, this technique showed itself to be excellent for this task. In this sense, our experience in terms of this kind of study will be commented.

Reference [79] used the combination of 2-D PAGE and mass spectrometry for obtaining a proteomic map for transgenic soybean seeds. The literature reports [12] that the number of protein spots present in the linear pH range from 4 to 7 was higher than the number of spots present in the linear range from 3 to 10 for transgenic soybean seeds. Therefore, for the range from 4 to 7, a higher number of spots were selected when compared with the 3-10 range. As result, a total of 192 proteins from transgenic soybean seeds were identified, 179 of them identified within the pH range from 4 to 7, and 13 of them identified within the 3-10 pH range. Regarding the pH range from 4 to 7, 49% of the spots present in the gel were identified in the database showing good efficiency with a similar study involving soybean published in the literature [80]. Regarding their biological functions, 50% were related to storage function, 18% related in growth/cell division process, 9% involved in metabolic/ energy process, 6% related to protein transport, 4% corresponding to proteins involved in the disease/defense category and 21% in the category of non-classified proteins.

The application of the 2-D PAGE technique in comparative proteomic studies can lead to some problems due to the intrinsic characteristics of the electrophoretic systems such as sample preparation strategies, the natural variations when considering biological systems, gel-to-gel variance, labor intensiveness and possible identification of several proteins from one spot [67,81]. In this sense, 2-DE technologies need to be evaluated critically.

In a pioneering work, reference [66] evaluates some parameters that influence the comparisons of the protein map after different gel runs, establishing comparative image analysis after 2-D PAGE of transgenic and non-transgenic soybean seeds for identifying possible differences in protein expression. In that work, two pH ranges were used: 3-10 and 4-7. For improving accuracy, image treatments were made by the same analyst and concomitantly carried out for each pair of gels in the same electrophoretic run (4 pairs of gels with the optimized loaded mass) for avoiding possible variations between evaluations of the gel images.

In relation to detection and selection of the protein spots, the choice of the parameters of image analysis is extremely important. Differences between manual and automatic detection of the spots were obtained, showing the importance in editing the images to avoid erroneous interpretations not only in terms of the quantities of the detected spots, but also in terms of the intensities and/or volume of each protein detected.

The matching study is of utmost importance for those ones where the target is to find possible changes in protein expression as well as to establish the similarities in protein distribution in sets of gels. For gels obtained in the same run and within 3 to 10 and 4 to 7 pH ranges, 163 ± 37 ($79\pm4\%$ match) and 287 ± 28 spots ($77\pm2\%$ match) were respectively obtained from 4 pairs of gels (transgenic x non-transgenic). However, when gels were obtained from different runs, even considering the same sample (transgenic seeds), high variation was detected in terms of matches ($39\pm6\%$ and $58\pm13\%$ for 3–10 and 4–7 pH ranges gels, respectively). Similar results for non-transgenic seeds were obtained ($40\pm10\%$ and $62\pm18\%$ for 3–10 and 4–7 pH gel ranges, respectively). In this way, it is preferable to acquire the gels in the same run in order to produce high matches. The use of these procedures points out that elimination of gel-to-gel variation is mandatory in image analysis.

Proteins were considered as up or down regulated when the ratio between spot volume and/or intensity for non-transgenic and transgenic soybean seeds changed from < 0.55 to >1.8 (*ca.* 90% variation). Thus, 3 and 7 spots from 3 to 10 and 4 to 7 pH ranges were respectively highlighted and characterized by MALDI-QTOF MS. From this total, 8 proteins were identified as: glycinin G2/A2B1 precursor, glycinin G1 precursor, α-subunit of β-conglycinin (03 spots), allergen Gly mBd 28 K (fragment), actin (fragment) and sucrose binding protein.

Then, it is easy to observe that well optimized conditions for acquiring images from 2D gels are an important tool in the identification of possible biomarkers for genetically modified organisms.

6. Two-dimensional difference gel electrophoresis: 2-D DIGE

A promising alternative for circumventing possible variations in the technique already described (2-D PAGE) is the two-dimensional difference gel electrophoresis (2-D DIGE). This technique, which is based on fluorescent cyanine dyes, allows comparisons between two exact quantitative proteomic samples, which are resolved on the same gel, minimizing the problems previously mentioned [82]. Moreover, there is the advantage of the high sensitivity of these dyes (*ca.* 1 fmol of protein), which enables the detection of low abundance proteins when compared to other dyes used in the detection of protein spots, such as Coomassie Brilliant Blue (CBB) and silver staining [82]. Frequently, three samples are labeled in 2-D DIGE: two of them are experimental samples whereas the third one is composed of a mixture of equal amounts of all experimental samples (*i.e.*, a pooled internal standard). This creates a standard for each protein during analysis. After 2D separation, different protein samples labeled can be visualized separately by exciting the different dyes at their specific excitation wavelengths. Therefore, from the images generated for each dye, the signals from labeled protein spots are determined and the normalized intensities or spot volumes for each spot from different dyes (Cy2, Cy3, Cy5) are compared in order to identify differentially expressed proteins between the samples [82-83].

Once this technique is finely developed for finding possible biomarkers, reference [79] applied the 2-D DIGE technique and mass spectrometry (ESI-QTOF MS/MS) to assess differences in proteomic profiles of transgenic and non-transgenic soybean seeds. Three biological replicates were analyzed. A regulation factor of 1.5 (50% variation) was chosen as determined by the image analysis program and statistically significant differences in expression were determined ($p \leq 0.05$, according to the Student t test). The program of image analysis uses the automatic detection of the spots, and does not require any manual editing, either in adding or in altering the area defining the spots, in contrast to other programs for 2-D PAGE image.

As a result, a total of four proteins were differentially expressed between transgenic and non-transgenic soybean seeds, where two were overexpressed, being more highly expressed in transgenic soybean, and two were underexpressed, being less expressed in transgenic soybean. Thus, these four spots were selected for identification by mass spectrometry. As results, the spots were identified as: Actin (fragment) (*Glycine max*), involved in various types of cell motility, widely expressed in all eukaryotic cells and binds to ATP and other proteins [84]; cytosolic glutamine synthetase (*Cucumis melo*) (Figure 1a), considered as a ligand enzyme, being highly expressed in many types of roots, binds ATP molecules and is responsible for the primary assimilation of ammonia in all living organisms, participates in nitrogen fixation [84]; Glycinin subunit G1 (*Glycine max*), responsible for the nutritional, physicochemical, and physiological characteristics of soybean seeds [85] and Glycine-rich RNA-binding protein (*Glycine max*) (Figure 1b), involved in cellular response to environmental and developmental conditions [84]. It is noteworthy that the actin protein was also detected by Brandão et al. [66] working with 2-D PAGE and image analysis, and with the same sample.

The results obtained in reference [79], comments about some differential proteins found, establishing a relationship between oxidative stress (ROS production) and genetic modification. In this way, spectrophotometric enzymatic assays demonstrate that soybean transgenic seeds (for glyphosate resistance) exhibited higher activities for APX, CAT and GR enzymes compared to non-transgenic. Considering these results, the authors concluded that the oxidative stressful condition in transgenic seeds resulted in an increase of H_2O_2, which is probably controlled by the action of APX and CAT and even GR. Related to SOD, reference [79] showed the results for SOD activity in nondenaturing polyacrylamide gel electrophoresis, and it was possible to observe eight SOD isoenzymes detected in both transgenic and non-transgenic soybean seeds, one as Mn-SOD, two as Fe-SOD and five as Cu/Zn-SOD. The authors commented that the reduction in SOD activity in transgenic seeds was much more a result of a reduction in the Fe-SOD isoenzymes activities. Finally, and as a conclusion, the genetic modification itself might have induced extra ROS generation.

Proteins involved in the RNA processing and alternative splicing, RNA transport, messenger RNA (mRNA) translation, mRNA stability, and mRNA silencing mechanisms have been shown to be required for normal plant development and the responses of plants to altered environments [86-87]. In our case, just the glycine-rich RNA-binding protein was differentially found after DIGE analysis, and this protein correlated to ROS production according to different articles [88, 89]. As already mentioned, the cytosolic glutamine synthetase is involved in nitrogen fixation. Oxidative stress can also control the expression of nitrogen-metabolism genes as recently demonstrated [90], demonstrating that cytosolic glutamine synthetase can be altered because of the oxidative stress observed in the transgenic soybean line [90].

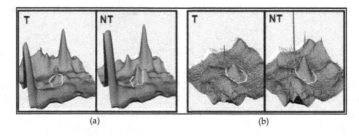

(a)   (b)

Figure 1. Examples of spots with expression variation among samples of transgenic (T) and non-transgenic (NT) soybean seeds: a - cytosolic glutamine synthetase (*Cucumis melo*) and b - Glycine-rich RNA-binding protein (*Glycine max*) [modified from reference 79].

7. Metallomics studies involving HPLC coupled to ICP-MS

In the last years, many soybean varieties have been genetically modified for adaptation to different geographical regions, to increase quality and productivity. Due to these genetic modifications, the proteins composition and profile can be affected, causing changes in the species proteome [91-92]. As previously described, the knowledge of the soybean genotype alone does not show enough information about the protein modifications due to environmental interactions. For better understanding of the consequences of a genetic manipulation, the elucidation of the protein map composition is necessary because it is directly related to the phenotype [93]. Since the proteome can be affected, it is assumed that the metallome can also be affected somehow by the genetic modification [94].

The metallome is defined as the entirety of metals and metalloid species, present in a cell or tissue type [95]. Deciphering the metallome provides information such as: (i) how an element is distributed among the cellular compartments; (ii) its coordination environment, in which the biomolecule is incorporated or by which bioligand it is complexed, and (iii) the concentration of the individual metal species present [96]. The majority of metals present in biological fluids and organs are linked to proteins, called metalloproteins. It is believed that every third protein require a metal cofactor, such as Cu, Fe and Zn, to develop their functions correctly [97]. The determination of an organism metallome involves separation techniques associated to microanalytic processes, such as mass spectrometry. These are the two key steps for general proteomics: separation and posterior identification of the proteins [80].

Metallomics studies were already performed in our group, being that one involving comparative metallomics of transgenic and non-transgenic soybean seeds [94], the first published in the literature. Soybean proteins were separated using two-dimensional polyacrylamide gel electrophoresis (2-D PAGE), tryptically digested, characterized using matrix assisted laser desorption ionization - quadrupole time of flight – mass spectrometry (MALDI-Q-TOF-MS) and mapped using synchrotron X-ray fluorescence radiation (SR-XRF). The following metallic ions were found: Ca(II), Cu(II), Fe(II), Mn(II), Ni(II) and Zn(II), and the quantitative profile was acquired using atomic absorption spectrometry, showing changes in metal contents of transgenic and non-transgenic soybean seeds. Although promising results could be found in this study, the canonical analytical approaches for proteomics (such as 2-D PAGE) and metabolomics studies usually do not consider the existence of metal complexes with proteins and metabolites.

In this way, the use of high performance liquid chromatography (HPLC), an analytical technique used to separate a mixture in solution in its individual components, should be considered. Distinctly compared from 2-D PAGE, HPLC is based on different protein-surface interactions [98]. The principal HPLC separation mechanisms used for bioinorganic studies include size exclusion chromatography (SEC), reversed-phase (RP) and ion-exchange (IEX) chromatography, and because of the complex nature of the metal-molecule interaction, a combination of these mechanisms is often necessary to identify the elemental species correctly.

Together with HPLC, an element-specific detector must be used, and since the 80's, inductively coupled plasma mass spectrometry (ICP-MS) is being widely applied for studying elements at low concentrations. It is robust for multielementar determinations, allowing to reach extremely low detection limits and giving isotopic information for identification and quantification of the species, besides being easily coupled to classic separation techniques, such as HPLC, readily realized since the chromatographic flow (0.5-1.0 mL min^{-1}) is compatible to common ICP-MS nebulizers [97,99].

The excitation source of this technique is argon inductively coupled plasma, which is used to form ions which are transferred to a high vacuum region through an interface containing small orifices. Ions are focalized using ionic lenses and directed to the mass spectrometer in order to be separated by m/z ratio. The m/z ratio analyzer generally used in ICP-MS is a quadrupole, ideal for quantitative analysis [100].

The elements detected using ICP-MS include metal coordination complexes with larger proteins and metallothioneins, as well as selenoproteins and metal/semi-metals linked to carbohydrates. A great amount of examples are found in the literature where ICP-MS is used to detect and quantify metallic ions bounded to biomolecules.

The attempts to avoid metal-ligand denaturation make the SEC mechanism the most used for metallomics studies [97]. It separates molecules according to their hydrodynamic volume, determined by their Stokes ratio. It results in the partial exclusion of analytes that pass through defined size pores due to their molecular sizes [101]. When the mobile phase passes through the column, those particles with small hydrodynamic volumes are transported through a larger path because they equilibrate in the pores more frequently than the ones with higher hydrodynamic volumes, resulting in separation. Elution volume is determined by the molecule size, directly related to their molecular weight, so this volume can be used to determine the molecular weight of an unknown compound. For carrying out that task, the relation between molecular weight and elution volume, obtained empirically by injecting standards with known molecular weight and measuring their elution volumes, must be known [102-103].

SEC is especially suitable for separation of element species presenting limited stability frequently found in protein-rich matrices. The main advantages of SEC are simplicity of application, tolerance to biological matrices, compatibility of mobile phases with specific demands of certain biological samples and the possibility of estimation of molecular weights of the compounds. It is widely used for protein separation, including soybean proteins and, although considered a low resolution method [104], it is often applied as the first separation method of fractions containing metallo-biomolecules of interest followed by another separation step with element-specific detection or MS identification. Its uses alone is very helpful to study the distribution of elements in different molecular weight fractions, and the coupling SEC-ICP-MS is being accepted as a hyphenated technique for speciation studies to evaluate the association of elements to compounds present in the sample. These studies are considered the initial point for a deeper evaluation of the nature of the species found.

In a work developed in our group [105], a comparison between elution profiles from transgenic and non-transgenic soybean seeds was carried out, using SEC coupled to high resolution ICP-MS. The elution profiles were similar between the samples, and the conclusions are in agreement to the ones discussed by reference [106]. It was found that areas of the most abundant peaks for Cu and Fe in transgenic soybean seeds were 3- and 2-fold higher, respectively, than those found in non-transgenic samples. This, summed up with total element analysis results in the same article, where the concentrations of Cu and Fe had statistically significant differences between transgenic and non-transgenic soybean seeds, could lead to the conclusion that Cu and Fe are associated with compounds more expressed in transgenic soybean seeds.

Reference [107] used different parts of the soybean plants to analyze Se elution profile using SEC-ICP-MS. The authors concluded that the bean had the most interesting profile, since it absorbed most of the Se from the shoots and presented a very intense peak for this element at higher molecular weight fraction. These data showed that the soybean plants convert selenite (used to enrich the plant) to high molecular weight species, which, according to the authors, can add nutritional value to the plant. Another work from the same group [108] used reversed phase coupled to ICP-MS to study the Se-Hg antagonism, and they found that in plants enriched with Hg, more Se was assimilated, indicating a possible protective response mechanism to the Hg.

As already commented, it is known that purity of peaks in SEC is poor, and even if a single species of a given element is present, matrix components may co-elute. They are invisible to the element specific detector, but if the goal is the identification of the organic specie linked to the element, they will be detected by the organic MS instrument. Also, matching the elution volume with a standard in this case is not definitive, due to the small number of theoretical plates found in SEC. For these reasons, SEC is usually followed by a second chromatographic separation (2nd dimension) using an orthogonal separation mechanism, such as ion-exchange, reversed-phase or hydrophilic interaction chromatography, before the identification of the components.

Multidimensional liquid chromatography is an efficient tool and an alternative procedure for the classic methods based on unidimensional HPLC. The multidimensional chromatography can be carried out *online* or *off-line*. In the *off-line* mode, fractions eluted from the 1st dimension are collected manually or using a fraction collector, and then are re-injected in the second chromatographic column. *Online* techniques are automated using a selector valve, which can enhance reliability and sample processing. The limitation here is that the mobile phases used for both dimensions must be compatible [109].

The selectivity in a multidimensional system can be enhanced only if the chromatographic dimensions are based in different separation mechanisms. The second dimension must not decrease the resolution obtained using the previous one. For the separation mechanisms to be different, the columns must have different stationary phases, allowing the less efficient separation attained in the first dimension to be improved in the second [110].

In the case of multidimensional liquid chromatography [103,111] coupled to ICP-MS, fractions isolated mainly using SEC, can be fractionated again using an independent separation mechanism to provide more detailed results, and also to attain metal species pure enough to be characterized using molecular mass spectrometry.

Many stationary phases can be used for a second chromatographic dimension. Among then, reversed phase (RP) [111], the most popular liquid chromatography separation mechanism, should be highlighted. It has great efficiency and is able to separate a great range of compounds with different polarities. The separation is obtained through partition of the analyte between a non-polar stationary phase and a polar mobile phase.

Ion exchange chromatography [112] (IEX) can also be used to separate biomolecules based in charge differences. It can be considered a highly selective technique, able to separate, for example, proteins differing in only one charged group. It is a widely used technique in bioseparations, since peptides, proteins, nucleic acids and related biopolymers have ionizable chemical domains, making them susceptible to enhancement or diminishment of their charges as a function of pH and ionic strength changes. It can be used to separate large biomolecules, with more than 60 kDa.

Finally, polar compounds can be efficiently separated using polar/hydrophilic stationary phases using normal phase aqueous chromatography (aqNPC), also called hydrophilic interaction chromatography (HILIC) [113]. Here, retention times tend to be longer as high as is the hydrophobicity of the solutes, indicating potential for small metallic complexes separation.

Concerning multidimensional chromatographic separations and soybeans, a recent work from reference [107] used IEX as second dimension for the separation of proteins from selenium-enriched soybean. Here, the target was only Se, and the 26 fractions collected from the second dimension were pure enough to allow the identification of a considerable number of proteins in the soybean databank.

In our group [114], IEX was also used as second dimension, generating a number of different fractions for both transgenic and non-transgenic soybean seeds used in that research. Taking as an example cobalt, the SEC separation (Figure 2a) provided 3 peaks, divided into F1, F2, and F3. When F3 is separated again using IEX (Figure 2b), the wide peak found using SEC was separated into two narrow peaks, showing that the separation resolution increased.

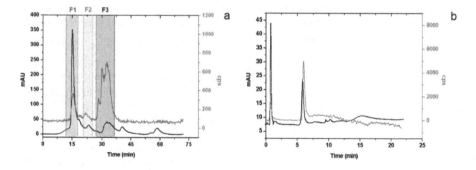

Figure 2. Chromatograms for UV absorption at 280 nm (—) and Co signal in the ICP-MS (—) for (a) SEC, separated in three fractions: F1, F2 and F3; and (b) IEX separation of F3 using transgenic soybean seeds [modified from supplementary material of reference 114].

Nowadays, mass spectrometry associated to bioinformatics has become essential in studies involving proteins, not only due to their sensitivity, but also to the total of information that can be obtained [69]. Electrospray ionization (ESI) is the most used technique for protein identification, allowing the formation of ions in the gas phase using a soft ionization process, making possible the analysis of non-volatile and thermolabile compounds [115]. As a consequence, ESI facilitated the analysis of large biomolecules, as well as drugs and their metabolites.

To improve metallomics information concerning transgenic and non-transgenic soybean seeds, our group [114] used the multidimensional chromatography strategy, as pointed out previously in this text. Total amounts of Fe and Cu were already found to be higher in transgenic soybeans, and in an attempt to link these metals to proteins, an ESI-MS/MS analysis was carried out. As results, more than 20 proteins were identified, encompassing 4 different functional categories. Among them, β-conglycinin, a protein previously associated to metals, was identified in three fractions, and one metalloprotein that binds Fe, lypoxigenase 1, was found in a high molecular weight fraction, the only fraction where an Fe peak was separated.

8. Future trends

Currently, the comparative studies concerning alterations in proteins, metalloproteins, metals and enzymes have demonstrated significant differences among transgenic and non-transgenic soybean. These differences have indicated that these genetic modifications provide not only tolerance to herbicide but also cause many changes in the whole metabolism of the transgenic plants.

Carefully taking into account all the results presented, it is possible to raise the following question: what are the future trends in comparative studies involving transgenic and non-transgenic soybean? Since the whole metabolism of transgenic soybean plants seems to be different to the non-transgenic one, this is a promising research area, and too much work is still needed. Much more information is still ahead of us for a better comprehension of the specific aspects of the transgenic soybean plant metabolism. In this way, investigations into techniques and novel approaches, quantitative proteomics, imaging and mapping of elemental distribution, tracer experiments employing stable isotopes and also in natural variation in the isotopic composition of the elements may possibly be the future trends in this topic. This will contribute to elucidation and expansion of our knowledge about transgenic soybean.

Since the proper functioning of life depends on the elements in a variety of processes, the understanding of molecular mechanisms of the elements and information on its chemical forms present in a living organism are very important. In this context, studies about identification and/or quantification of one or more chemical species of elements in transgenic and non-transgenic soybean samples are able to generate valuable information about their metabolisms. Therefore, it would be useful if more efforts were devoted to this topic. A novel technique that has an unexplored potential for speciation analysis is travelling wave ion mobility spectrometry coupled to mass spectrometry (TWIMS-MS) [116]. The use of this technique in speciation analysis of metals associated with biomolecules should increase due to its capability of differentiation of ions by shape and size, besides mass and charge. Until now, the studies employing ion mobility are concentrated in proof-of-the-concept using isolated species commercially available and its application to complex matrices certainly will be a big challenge, but very helpful to elucidate many questions.

The main objective of quantitative proteomics is to quantify protein expression alterations in response to a variety of changes, and, nowadays, one of the most challenging and emerging area of proteomics involves the developments of accurate quantitative methods for proteins. The quantitative proteomics is divided in absolute and relative subjects. In the absolute quantification, changes in protein expression are determined in exact amount or concentration of each protein present. The relative one determines the up- or down-regulation of a protein relative to the control sample, and the results are presented as 'fold' increases or decreases. The 2-D DIGE is an example of relative quantification technique that is applied to intact proteins and the differential expression determination is based on fluorescence as commented earlier in this chapter. Taking into account the soybean comparative studies the application of quantitative proteomics by 2-D DIGE or by other technique could continue establishing the differences in protein expressions accurately [117].

According to the results presented earlier, some elements are present at higher concentrations in transgenic soybean seeds than in non-transgenic ones [66,94,105]. The transgenic seed seems to have ability to take up higher amounts of some metals from the soil and this is a sign that the processes involved in intake, transport and storage of essential and toxic metals and metalloids probably are suffering changes due to genetic modification. Various new queries take place with this information, such as: The other transgenic plant parts (roots, stems and leaves), are also taking up, transporting and storing higher amounts of these metals? Other plant parts try to eliminate some excess of these metals? Are these higher amounts really an excess for a transgenic plant or not? Are there differences in the distributions of these metals among transgenic and non-transgenic soybean? A potential tool for obtaining a better insight in these processes can be to use tracer experiments employing stable isotopes. In the last few years the use of stable isotopes and their isotope ratio measurements have gained importance for tracer experiments in biological and medical research [118]. In these studies stable isotopic tracers with an isotopic composition sufficiently different from the corresponding natural one is added to the studied system and changes in the selected isotope ratio monitored. The absorption or bioavailability of an element can be determined with this approach as well as information about element redistribution over various compartments of an organism [118-119]. According to our knowledge, no tracer study for essential or toxic metals evaluating transgenic and non-transgenic soybean is found in the literature and therefore there is a great amount of work to perform in this challenging area.

LA-ICP-MS offers *in situ* analysis of solid samples with respect to metals and nonmetals at trace concentration level mostly without sample preparation and without charging effects during the measurements. This technique can also be applied to the imaging of soft tissues such as plant leaves with relatively high spatial resolution and good sensitivity [120] and therefore, some investigations involving metals distribution by LA-ICP-MS in transgenic and non-transgenic soybean would also be a future trend.

Another challenging issue that can provide evidences supporting the hypothesis that genetic modification is affecting the metabolism of soybean plants involves the investigation of natural variation in the isotopic composition of the elements. Even though isotopic abundances are assumed to be almost constant in nature, small isotopic or mass fractionation effects occur in both natural and industrial processes [118]. Since the isotopes present the same number of electrons, they show basically the same chemical behavior. However, there is a small discrepancy in their physicochemical behavior due to the mass difference, which may leave isotopes of the same element to take part with slightly different efficiencies in physical processes or (bio)chemical reactions, and, consequently, to result in variations in the isotopic composition [118-119]. These differences in efficiency are associated to a minor distinction in equilibrium for each different isotopic molecule - thermodynamic effect or in the rate with which the isotopes participate in a process or reaction - kinetic effect. Lighter elements, such as H, C, N, O and S suffer more pronounced isotopic variations because of the high relative mass difference between their isotopes. Nevertheless, heavier elements are subject to isotope fractionation, even though the change is minor [121].

As relative abundances cannot be measured directly, these studies are based on measuring the isotope ratio of an element because it is experimental accessible. The isotope ratio measured in a particular sample (R_x) is compared to the corresponding one in another sample, frequently a reference sample (R_{RF}) [118]. The differences found are frequently very small and thus high reproducibility/repeatability is required. Thus, the ICP-MS technique is becoming the more advantageous choice for most applications employing isotope ratios, mainly considering the recent instrumental developments. As the elements are subject to isotope fractionation in nature, the genetic modification could also provoke or intensify this effect.

In view of that comment here, it is easy to rationalize that many aspects can be explored when focusing on studies related to transgenic soybean.

9. Conclusion

The initial hypothesis formulated that the genetic modification itself is stressing the soybean, is apparently right, once the plant is searching a new equilibrium as living organism. The results presented in this chapter demonstrate that not only is the proteomic map changed with some proteins increasing and others decreasing, but also chromatographic separations are altered when transgenic and non-transgenic soybeans are compared. Examples are activities of some enzymes (as CAT, SOD, GPx, among others) involved in neutralization of ROS, as well as the possible capacity in taking metals from the soil (mainly for Fe and Cu). Because of these modifications that occur when both transgenic and non-transgenic organisms are compared, the theme of genetic modification could be even better explained with some alternative strategies, such as quantitative proteomics, image analysis, tracer experiments with stable isotopes, and other possibilities.

Finally, in our point of view, one of the key points for the success of studies involving transgenic organisms is not only to involve good technology, but also a transdisciplinary view, involving different areas of expertise. With this strategy, it will be easier to understand this area of investigation, making possible the demystification of the genetic modification that have occurred, and allowing answers for some questions that still remain unknown.

10. Nomenclature and acronyms

2-D-HPLC Two-Dimensional High Performance Liquid Chromatography

2-D PAGE Two-Dimensional Gel Electrophoresis

2-D DIGE Two-Dimensional Difference Gel Electrophoresis

ICP-MS Inductively Coupled Plasma Mass Spectrometry

HR-SF-ICP-MSHighResolutionSectorFieldInductivelyCoupledPlasmaMassSpectrometry

LA-ICP-MS Laser Ablation Inductively Coupled Plasma Mass Spectrometry

MALDI-QTOF-MS Matrix-Assisted Laser Desorption Ionisation Quadrupole-Time-of-Flight Mass Spectrometry

ESI-LC-MS-MS ElectroSpray Ionization Liquid Chromatography Mass Spectrometry-Mass Spectrometry

2-D-HPLC-ICP-MS Two-Dimensional High Performance Liquid Chromatography Inductively Coupled Plasma Mass Spectrometry

SDS-PAGE Sodium Dodecyl Sulfate Polyacrylamide Gel Electrophoresis

ESI-QTOFMS-MSElectroSprayIonization-Time-of-FlightMassSpectrometry-MassSpectrometry

SEC-ICP-MSSizeExclusionChromatography-InductivelyCoupledPlasmaMassSpectrometry

Acknowledgements

The authors are grateful to the Fundação de Amparo a Pesquisa do Estado de São Paulo (FAPESP, São Paulo, Brazil), Conselho Nacional de Desenvolvimento Científico e Tecnológico (CNPq, Brasília, Brazil), and Coordenação de Aperfeiçoamento de Pessoal de Nível Superior (CAPES, Brasília, Brazil), for financial support.

Author details

Marco Aurélio Zezzi Arruda[1], Ricardo Antunes Azevedo[2], Herbert de Sousa Barbosa[1], Lidiane Raquel Verola Mataveli[1], Silvana Ruella Oliveira[1], Sandra Cristina Capaldi Arruda[2] and Priscila Lupino Gratão[3]

1 Institute of Chemistry, National Institute of Science and Technology for Bioanalytics and Department of Analytical Chemistry, University of Campinas, Campinas, Brazil

2 Department of Genetics, Laboratory of Genetics Biochemistry of Plants, University of São Paulo, ESALQ, Piracicaba, Brazil

3 Department of Applied Biology to Agricultural, Universidade Estadual Paulista, FCAV, Jaboticabal, Brazil

References

[1] Sharma S, Kaur M, Goysil R, Gil BS. Physical Characteristics and Nutritional Composition of Some New Soybean (Glycine max (L.) Merrill) genotypes. Journal of Food Science and Technology 2011; DOI:10.1007/s13197-011-0517-7.

[2] The American Soybean Association. SoyStats. http://www.soystats.com (accessed 3 May 2012).

[3] Rural Centro. http://www.ruralcentro.com.br/noticias/53562/usda-reduz-producao-de-soja-da-safra-201112-para-2515-milhoes (accessed 21 March 2012).

[4] Cavusoglu K, Yapar K, Oruc E, Yalcin E. Protective Effect of Ginkgo Biloba L. Leaf Extract Against Glyphosate Toxicity in Swiss Albino Mice. Journal of Medicinal Food 2011;14(10) 1263-1272.

[5] Windels P., Tavernies I., Depicker A., Van Bockstaele E., De Loose M. Characterisation of the Roundup Ready Soybean Insert. European Food Research and Technology 2001;213(2) 107-112.

[6] Monsanto do Brasil. http://www.monsanto.com.br (acessed May 15 2012).

[7] Saz JM, Marina ML. High Performance Liquid Chromatography and Capillary Electrophoresis in the Analysis of Soybean Proteins and Peptides in Foodstuffs. Journal of Separation Science 2007;30(4) 431-451.

[8] Scandalios JG. Oxidative Stress: Molecular Perception and Transduction of Signals Triggering Antioxidant Gene defenses. Brazilian Journal of Medical and Biological Research 2005;38(7) 995-1014.

[9] Scandalios J.G., Guan L., Polidoros A.N. Catalases in Plants:Gene Structure, Properties, Regulation, and Expression. In: Scandalios J.G. (ed.) Oxidative Stress and the Molecular Biology of Antioxidant Defenses. New York: Cold Spring Harbor Laboratory Press Plainview; 1997. p343-406.

[10] Fridovich I. Superoxide Radical and Superoxide Dismutases. Annual Review of Biochemistry 1995;64 87-112.

[11] Hippeli S, Heiser I, Elstner EF. Activated Oxygen and Free Oxygen Radicals in Pathology:New Insights and Analogies Between Animals and Plants. Plant Physiology Biochemistry 1999;37(3) 167-178.

[12] Levine A, Tenhaken R, Dixon R, Lamb C. H2O2 from the Oxidative Burst Orchestrates the Plant Hypersensitive Disease Resistance Response. Cell 1994;79(4) 583-593.

[13] Dat J, Vandenabeele S, Vranová E, Van Montagu M, Inzé D, Van Breusegem F. Dual Action of the Active Oxygen Species During Plant Stress Responses. Cellular and Molecular Life Sciences 2000;57(5) 779-795.

[14] Foyer, CH, Shigeru S. Understanding Oxidative Stress and Antioxidant Functions to Enhance Photosynthesis. Plant Physiology 2011;155(1) 93-100.

[15] Foyer, CH, Noctor G. Redox Homeostasis and Antioxidant Signaling: A Metabolic Interface between Stress Perception and Physiological Responses. The Plant Cell 2005;17(7) 1866-1875

[16] Gratão PL, Polle A, Lea PJ, Azevedo RA. Making the Life of Heavy Metal-Stressed Plants a Little Easier. Functional Plant Biology 2005;32(6) 481-494.

[17] Jones DP. Redefining Oxidative Stress. Antioxidant Redox Signaling 2006;8(9-10) 1865-1879.

[18] Foyer CH, Noctor G. Oxidant and Antioxidant Signaling in Plants: a Re-Evaluation of the Concept of Oxidative Stress in a Physiological Context. Plant Cell Environment 2005;28(8) 1056-1071.

[19] Foyer CH, Noctor G. Oxygen Processing in Photosynthesis: Regulation and Signaling. New phytologist 2000;146(3) 359-388.

[20] Rizhsky L, Hallak-Herr E, Van Breusegem F, Rachmilevitch S, Barr JE, Rodermel S, Inzé D, Mitler R. Double Antisense Plants Lacking Ascorbate Peroxidase and Catalase are Less Sensitive to Oxidative Stress than Single Antisense Plants Lacking Ascorbate Peroxidase or Catalase. Plant Journal 2002;32(3) 329-342.

[21] Moller IM. Plant Mitochondria and Oxidative Stress: Electron Transport, NADPH Turnover, and Metabolism of Reactive Oxygen Species. Annual Review of Plant Physiology and Plant Molecular Biology 2001;52 561-591.

[22] Medici LO, Azevedo RA, Smith RJ, Lea PJ. The Influence of Nitrogen Supply on Antioxidant Enzymes in Plant Roots. Functional Plant Biology 2004;31(1) 1-9.

[23] Bechtold U, Murphy DJ, Mullineaux PM. Arabidopsis Peptide Methionine Sulfoxide Reductase 2 Prevents Cellular Oxidative Damage in Long Nights. The Plant Cell 200;16(4) 908-919.

[24] Wojtaszek P. Oxidative Burst: an Early Plant Response to Pathogen Infection. Biochemical Journal 1997;322(3) 681-692.

[25] Yu W, Zhang R, Runzhi L, Sandui G. Isolation and Characterization of Glyphosate-Regulated Genes in Soybean Seedlings. Plant Science 2007;172(3) 497-504.

[26] Moldes CA, Médici LO, Abrahão OS, Tsai SM, Azevedo RA. Biochemical Responses of Glyphosate Resistant and Susceptible Soybean Plants Exposed to Glyphosate. Acta Physiologiae Plantarum 2008;30(4) 469-479.

[27] Knörzer O, Burner J, Boger P. Alterations in the Antioxidative System of Suspension-Cultured Soybean Cells (Glycine Max) Induced by Oxidative Stress. Physiologia Plantarum 2008;97(2) 388-396.

[28] Gill SS, Nafees AK, Narendra T. Differential Cadmium Stress Tolerance in Five Indian Mustard (Brassica Juncea L.) Cultivars. An evaluation of the role of antioxidant machinery. Plant Signaling and Behavior 2011;6(2) 293-300.

[29] Mittler R, Blumwald E. Genetic Engineering for Modern Agriculture: Challenges and Perspectives. Annual Review in Plant Biology. 2010;61 443-462

[30] Ghelfi A, Gaziola SA, Cia MC, Chabregas SM, Falco MC, Kuser-Falcão PR, Azevedo RA. Cloning, Expression, Molecular Modeling and Docking Analysis of Glutathione Transferase from Saccharum Officinarum. Annals of Applied Biology 2011;159(2) 267-280.

[31] Delaunay A, Pflieger D, Barrault MB, Vinh J, Toledano MB. A Thiol Peroxidase is an H2O2 Receptor and Redox-Transducer in Gene Activation. Cell 2002;111(4) 471-481.

[32] Arruda MAZ., Azevedo RA. Metallomics and Chemical Speciation: Towards a Better Understanding of Metal-Induced Stress in Plants. Annals of Applied Biology 2009;155(3) 301-307.

[33] Apel K, Hirt H. Reactive Oxygen Species: Metabolism, Oxidative Stress and Signal Transduction. Annual Review on Plant Biology 2004;55 373-399.

[34] Galant A, Koester RP, Ainsworth EA, Hicks LM. From climate change to molecular response: redox proteomics of ozone-induced responses in soybean. New Phytologist 2012;194(1) 220-229.

[35] Zhao T, Wang JL, Wang Y, Cao Y. Effects Of Antioxidant Enzymes of Ascorbate-Glutathione Cycle in Soybean (Glycine Max) Leaves Exposed to Ozone. Advanced Research on Industry, Information Systems and Material Engineering, pts 1-7 Advanced Materials Research 2011;204-210(1-7) 672-677.

[36] Gillespie KM, Rogers A, Ainsworth EA. Growth at Elevated Ozone or Elevated Carbon Dioxide Concentration Alters Antioxidant Capacity and Response to Acute Oxidative Stress in Soybean (Glycine Max). Journal of Experimental Botany 2011;62(8) 2667-2678.

[37] Singh E, Tiwari S, Agrawal M. Variability in Antioxidant and Metabolite Levels, Growth and Yield of Two Soybean Varieties: an Assessment of Anticipated Yield Losses under Projected Elevation of Ozone. Agriculturall Ecosystems & Environment 2010;135(3) 168-177.

[38] Masoumi H, Darvish F, Daneshian J, Normohammadi G, Habibi D. Effects of Water Deficit Stress on Seed Yield and Antioxidants Content in Soybean (Glycine Max L.) Cultivars. African Journal of Agricultural Research 2011;6(5) 1209-1218.

[39] Masoumi H, Darvish F, Daneshian J, Nourmohammadi G, Habibi D. Chemical and Biochemical Responses of Soybean (Glycine Max L.) Cultivars to Water Deficit Stress. Australian Journal of Crop Science 2011;5(5) 544-553.

[40] Dogan M. Antioxidative and Proline Potentials as a Protective Mechanism in Soybean Plants under Salinity Stress. African Journal of Biotechnology. 2011;10(32) 5972-5977.

[41] Moussa HR. Low Dose of Gamma Irradiation Enhanced Drought Tolerance in Soybean. Bulgarian Journal of Agricultural Science 2011;17(1) 63-72.

[42] Kocsy G, Laurie R, Szalai G, Szilagyi V, Simon-Sarkadi L, Galiba G, de Ronde JA. Genetic Manipulation of Proline Levels Affects Antioxidants in Soybean Subjected to Simultaneous Drought and Heat Stresses. Physiologia Plantarum 2005;124(2) 227-235.

[43] Yi H, Ravilious GE, Galant A, Krishnan HB, Jez JM. From Sulfur to Homoglutathione: Thiol Metabolism in Soybean. Amino Acids 2010;39(4) 963-978.

[44] Kim WS, Chronis D, Juergens M, Schroeder AC, Hyun SW, Jez JM, Krishnan HB. Transgenic Soybean Plants Overexpressing O-Acetylserine Sulfhydrylase Accumulate Enhanced Levels of Cysteine and Bowman-Birk Protease Inhibitor in Seeds. Planta 2012;235(1) 13-23.

[45] Kumar V, Rani A, Dixit AK, Bhatnagar D, Chauhan GS. Relative Changes in Tocopherols, Isoflavones, Total Phenolic Content, and Antioxidative Activity in Soybean Seeds at Different Reproductive Stages. Journal of Agricultural and Food Chemistry 2009;57(7) 2705-2710.

[46] Pawlak S, Firych A, Rymer K, Deckert J. Cu,Zn-Superoxide Dismutase is Differently Regulated by Cadmium and Lead in Roots of Soybean Seedlings. Acta Physiologiae Plantarum 2009;31(4) 741-747.

[47] Melo LCA, Alleoni LRF, Carvalho G, Azevedo RA. Cadmium- and Barium-Toxicity Effects on Growth and Antioxidant Capacity of Soybean (Glycine Max L.) Plants, Grown in Two Soil Types with Different Physicochemical Properties. Journal of Plant Nutrition and Soil Science 2011;174(5) 847-859.

[48] Liu TT, Wu P, Wang LH, Zhou Q. Response of Soybean Seed Germination to Cadmium and Acid Rain. Biological Trace Element Research 2011;144(1-3) 1186-1196.

[49] Ershova AN, Popova NV, Berdnikova OS. Production of Reactive Oxygen Species and Antioxidant Enzymes of Pea and Soybean Plants under Hypoxia and High CO2 Concentration in Medium. Russian Journal of Plant Physiology 2011;58(6) 982-990.

[50] Shamsi IH, Wei K, Zhang GP, Jilani GH, Hassan MJ. Interactive Effects of Cadmium and Aluminum on Growth and Antioxidative Enzymes in Soybean. Biologia Plantarum 2008;52(1) 165-169.

[51] Sousa NR, Ramos MA, Marques APGC, Castro PML. The Effect of Ectomycorrhizal Fungi Forming Symbiosis with Pinus Pinaster Seedlings Exposed to Cadmium. Science of the Total Environment 2012;414 63-67.

[52]   Bressano M, Curetti M, Giachero L, Gil SV,Cabello M, March G, Ducasse DA, Luna CM. Mycorrhizal Fungi Symbiosis as a Strategy against Oxidative Stress in Soybean Plants. Journal of Plant Physiology 2010;167(18) 1622-1626.

[53]   Ghorbanli M, Ebrahimzadeh H, Sharifi M. Effects of Nacl and Mycorrhizal Fungi on Antioxidative Enzymes in Soybean. Biologia Plantarum 2004;48(4) 575-581.

[54]   Sugiyama M, Ae N, Arao T. Role of Roots in Differences in Seed Cadmium Concentration among Soybean Cultivars - Proof by Grafting Experiment. Plant Soil 2007;295(1-2) 1–11.

[55]   Fields S. Proteomics in Genomeland. Science, 2001;291(5507) 1221-1224.

[56]   Kersten B, Bürkle L, Kuhn EJ, Giavalisco P, Konthur Z, Lucking A, Walter G, Eickhoff H, Schneider U. Large-Scale Plant Proteomics. Plant Molecular Biology 2002;48 133-141.

[57]   Job D, Haynes PA, Zivy M. Plant Proteomics. Proteomics 2011;11(9) 1557-1558.

[58]   Wang Y, Kim SG, Kim ST, Agrawal GK, Rakwal R, Kang KY. Biotic Stress-Responsive Rice Proteome: an Overview. Journal of Plant Biology 2011;54(4) 219-226.

[59]   Hossain Z, Nouri MZ, Komatsu S. Plant Cell Organelle Proteomics in Response to Abiotic Stress. Journal of Proteome Research 2012;11(1) 37-48.

[60]   Komatsu S, Konishi H, Hashimoto M. The Proteomics of Plant Cell Membranes. Journal of Experimental Botany 2007;58(1) 103-112.

[61]   Ito J, Batth TS, Petzold CJ, Redding-Johanson AM, Mukhopadhyay A, Verboom R, Meyer EH, Millar AH, Heazlewood JL. Analysis of the Arabidopsis Cytosolic Proteome Highlights Subcellular Partitioning of Central Plant Metabolism. Journal of Proteome Research 2011;10(4) 1571-1582.

[62]   Cortleven A, Noben JP, Valcke R. Analysis of the Photosynthetic Apparatus in Transgenic Tobacco Plants with Altered Endogenous Cytokinin Content: A Proteomic Study. Proteome Science 2011;9(33) 1-14.

[63]   Satoh R, Nakamura R, Komatsu A, Oshima M, Teshima R. Proteomic Analysis of Known and Candidate Rice Allergens between Non-Transgenic and Transgenic Plants. Regulatory Toxicology and Pharmacology 2011;59(3) 437-444.

[64]   Uzogara SG. The Impact of Genetic Modification of Human Foods in the 21st Century: a Review. Biotechnology Advances 2000;18(3) 179-206.

[65]   Kim Y, Choi SJ, Lee H, Moon TW. Quantitation of CP4 5-Enolpyruvylshikimate-3-Phosphate Synthase in Soybean by Two-Dimensional Gel Electrophoresis. Journal of Microbiology and Biotechnology 2006;16(1) 25-31.

[66]   Brandão AR, Barbosa HS, Arruda MAZ. Image Analysis of Two-Dimensional Gel Electrophoresis for Comparative Proteomics of Transgenic and Non-Transgenic Soybean Seeds. Journal of Proteomics 2010;73(8) 1433-1440.

[67] Rabilloud T, Lelong C. Two-Dimensional Gel Electrophoresis in Proteomics: a Tutorial. Journal of Proteomics 2011;74(10) 1829-1841.

[68] Garcia JS, Souza GHMF, Eberlin MN, Arruda MAZ. Evaluation of Metal-Ion Stress in Sunflower (Helianthus Annus L.) Leaves through Proteomic Changes. Metallomics 2009;1(1) 107-113.

[69] Domon B., Aebersold R. Mass Spectrometry and Protein Analysis. Science 2006;312(5771) 212-217.

[70] Aebersold R, Mann M. Mass Spectrometry-Based Proteomics. Nature 2003;422(6928) 198-207.

[71] Sussulini A, Garcia JS, Mesko MF, Moraes DP, Flores EMM, Perez CA, Arruda MAZ. Evaluation of Soybean Seed Protein Extraction Focusing on Metalloprotein Analysis. Microchimica Acta 2007;158(1-2) 173-180.

[72] Arruda SCC, Barbosa HS, Azevedo RA, Arruda MAZ. Two-Dimensional Difference Gel Electrophoresis Applied for Analytical Proteomics: Fundamentals and Applications to the Study of Plant Proteomics. Analyst 2011;136(20) 4119-4126.

[73] Natarajan SS. Natural variability in abundance of prevalent soybean proteins. Regulatory Toxicology and Pharmacology 2010;58(3) 26-29.

[74] Cheng L, Gao X, Li S, Shi M, Javeed H, Jing X, Yang G, He G. Proteomic Analysis of Soybean [Glycine Max (L.) Meer.] Seeds During Inhibition at Chilling Temperature. Molecular Breeding 2010;26(1) 1-17.

[75] Toorchi M, Yukawa K, Nouri MZ, Komatsu S. Proteomics Approach for Identifying Osmotic-Stress-Related Proteins in Soybean Roots. Peptides 2009;30(12) 2108-2117.

[76] El-shemy HA, Khalafalla MM, Fujita K, Ishimoto M. Improvement of Protein Quality in Transgenic Soybean Plants. Biologia Platarum 2007;51(2) 277-284.

[77] Ocana MF, Fraser PD, Patel RKP, Halket JM, Bramley PM. Mass Spectrometric Detection of CP4 EPSPS in Genetically Modified Soya and Maize. Rapid Communication in Mass Spectrometry 2007;21(3) 319-328.

[78] Wittmann-Liebold B, Graack HR, Pohl T. Two-Dimensional Gel Electrophoresis as Tool for Proteomics Studies in Combination with Protein Identification by Mass Spectrometry. Proteomics 2006;6(17) 4688-4703.

[79] Barbosa HS, Arruda SCC, Azevedo RA, Arruda MAZ. New Insights on Proteomics of Transgenic Soybean Seeds: Evaluation of Differential Expressions of Enzymes and Proteins. Analytical and Bioanalytical Chemistry 2012;402(1) 299-314.

[80] Xu C, Garrett WM, Sullivan J, Caperna T, Natarajan SS. Separation and Identification of Soybean Leaf Proteins by Two-Dimensional Gel Electrophoresis and Mass Spectrometry. Phytochemistry 2006;67(22) 2431-2440.

[81] Rabilloud T, Chevallet M, Luche S, Lelong C. Two-Dimensional Gel Electrophoresis in Proteomics: Past, Present and Future. Journal of Proteomics 2010;73(11) 2064-2077.

[82] Timms JF, Cramer R. Difference Gel Electrophoresis. Proteomics, 2008;8(23-24) 4886-4897.

[83] Marouga R, David S, Hawkins E. The Development of the DIGE System: 2D Fluorescence Difference Gel Analysis Technology. Analytical and Bioanalytical Chemistry 2005;382(3) 669-678.

[84] Wang W, Vinocur B, Altman A. Plant Responses to Drought, Salinity and Extreme Temperatures: Towards Genetic Engineering for Stress Tolerance. Planta 2003;218(1) 1-14.

[85] Thanh VH, Shibasaki K. Major Proteins of Soybean Seeds. Subunit Structure of B-Conglycinin. Journal of Agricutural and Food Chemistry 1978;26(3) 692-695.

[86] Cheng Y, Chen X. Posttranscriptional Control of Plant Development. Current Opinion in Plant Biology 2004;7(1) 20-25.

[87] Wang QL, Li ZH. The Functions of Micrornas in Plants. Frontiers in Bioscience 2007;12 3975-3982.

[88] Schmidt F, Marnef A, Cheung MK, Wilson I, Hancock J, Staiger D, Ladomery M. A Proteomic Analysis of Oligo(Dt)-Bound Mrnp Containing Oxidative Stress-Induced Arabidopsis Thaliana RNA-Binding Proteins ATGRP7 and ATGRP8. Molecular Biology Reports 2010;37(2) 839-845.

[89] Kim YO, Pan S, Jung CH, Kang H. A Zinc Finger-Containing Glycine-Rich RNA-Binding Protein, Atrz-1a, has a Negative Impact on Seed Germination and Seedling Growth of Arabidopsis Thaliana under Salt or Drought Stress Conditions. Plant and Cell Physiology 2007;48(8)1170-1181.

[90] Pageau K, Reisdorf-Cren M, Morot-Gaudry JF, Masclaux-Daubresse C. The Two Senescence-Related Markers, GS1 (Cytosolic Glutamine Synthetase) and GDH (Glutamate Dehydrogenase), Involved in Nitrogen Mobilization, are Differentially Regulated during Pathogen Attack and by Stress Hormones and Reactive Oxygen Species in Nicotiana Tabacum L. Leaves. Journal of Experimental Biology 2006;57(3) 547-557.

[91] Funke T, Han H, Fried MLH, Fischer M, Schonbrünn E. Molecular Basis for the Herbicide Resistance of Roundup Ready Crops. Proceedings of the National Academy of Sciences of the United States of America 2006;103(35) 13010-13015.

[92] Natarajan S, Xu C, Caperna TJ, Garrett WM. Comparison of Protein Solubilization Methods Suitable for Proteomic Analysis of Soybean Seed Proteins. Analytical Biochemistry 2005;342 214-220.

[93] Natarajan S, Xu C, Bae H, Bailey BA, Cregan P, Caperna TJ, Garrett WM, Luthria D. Proteomic Analysis of Glycin Subunits of Sixteen Soybean Genotypes. Plant Physiology and Biochemistry 2007;45(6-7) 436-444.

[94] Sussulini A, Souza GHMF, Eberlin MN, Arruda MAZ. Comparative Metallomics for Transgenic and Non-Transgenic Soybeans. Journal of Analytical Atomic Spectrometry 2007;22(12) 1501-1506.

[95] Mounicou S, Szpunar J, Lobinski R. Metallomics: The Concept and Methodology. Chemical Society Reviews 2009;38(4) 1119-1138.

[96] Szpunar J. Advances in Analytical Methodology for Bioinorganic Speciation Analysis: Metallomics, Metalloproteomics and Heteroatom-Tagged Proteomics and Metabolomics. Analyst 2005;130(4) 442-465.

[97] Mounicou S, Lobinski R. Challenges to Metallomics and Analytical Chemistry Solutions. Pure and Applied Chemistry 2008;80(12) 2565-2575.

[98] Garcia JS, Magalhães CS, Arruda MAZ. Trends in Metal-Binding and Metalloproteins Analysis. Talanta 2006;69(1) 1-15.

[99] Sanz-Mendel A, Montes-Bayón M, Sánchez MLF. Trace Elements Speciation by ICP-MS in Large Biomolecules and its Potential for Proteomics. Analytical and Bioanalytical Chemistry 2003;377(2) 236-247.

[100] Thomas RA. A Beginners Guide to ICP-MS. Spectroscopy 2002;17(1) 36-41.

[101] Wagner K, Racaityte K, Unger KK, Miliotis T, Edholm LE, Bischoff R, Marko-Varga G. Protein Mapping by Two-Dimensional High Performance Liquid Chromatography. Journal of chromatography A 2000;893(2) 293-305.

[102] Neue UD. HPLC Columns: Theory, Technology and Practice. New York: Wiley-VCH Inc.; 1997.

[103] Silva MAO, Mataveli LRV, Arruda MAZ. Liquid Chromatography for Biosseparations: Fundamentals, Developments and Applications. Brazilian Journal of Analytical Chemistry 2011;1(5) 234-245.

[104] Wuilloud RG, Kannankumarath SS, Caruso JA. Speciation of Nickel, Copper, Zinc and Manganese in Different Edible Nuts: a Comparative Study of Molecular Size Distribution by SEC-UV-ICP-MS. Analytical and Bioanalytical Chemistry 2004;379(3) 495-503.

[105] Mataveli LRV, Pohl P, Mounicou S, Szpunar J. A Comparative Study of Elements Concentration and Binding in Transgenic and Non-Transgenic Soybeans Seeds. Metallomics 2010;2(12) 800-805.

[106] Koplik R, Pavelková H, Cincibuchová J, Mestek O, Kvasnicka F, Suchánek M. Fractionation of Phosphorus and Trace Elements Species in Soybean Flour and Common White Beans by Size Exclusion Chromatography-Inductively Coupled Plasma Mass Spectrometry. Journal of Chromatography B 2002;770(1-2) 261-273.

[107] Chan Q, Caruso JA. A Metallomics Approach Discovers Selenium-Containing Proteins in Selenium-Enriched Soybeans. Analytical and Bioanalytical Chemistry 2012;403(5) 1311-1321.

[108] Chan Q, Afton SE, Caruso JA. Selenium Speciation Profiles in Selenite-Enriched Soybean (Glycine Max) by HPLC-ICPMS and ESI-ITMS. Metallomics 2010;2(2) 147-153.

[109] dos Santos Neto AJ. Multidimensional Liquid Chromatography and Tandem Mass Spectrometry for the Direct Analysis of Drugs in Biofluids: from the Conventional to the Miniaturized Scale. PhD thesis. University of São Paulo; 2007.

[110] Stroink T, Ortiz MC, Bult A, Lingeman H, de Jong JG, Underberg WJM. On-Line Multidimensional Liquid Chromatography and Capillary Electrophoresis Systems for Peptides and Proteins. Journal of Chromatography B 2010;817(1) 449-466.

[111] Lobinski R, Schaumlöffel D, Szpunar J. Mass Spectrometry in Bioinorganic Analytical Chemistry. Mass Spectrometry Reviews 2006;25(2) 255-289.

[112] Ouerdane L, Mari S, Czernic P, Lebrun M, Lobinski R. Speciation of Non-Covalent Nickel Species in Plant Tissue Extracts by Electrospray Q-TOFMS/MS after their Isolation by 2D Size Exclusion-Hydrophilic Interaction LC (SEC-HILIC) Monitored by ICP-MS. Journal of Analytical Atomic Spectrometry 2006;21(7) 676-683.

[113] Mounicou S, Szpunar J, Lobinski R, Andrey D, Blake CJ. Bioavailability of Cadmium and Lead in Cocoa: Comparison of Extraction Procedures Prior to Size-Exclusion Fast-Flow Liquid Chromatography with Inductively Coupled Plasma Mass Spectrometric Detection (SEC-ICP-MS). Journal of Analytical Atomic Spectrometry 2002;17(8) 880-886.

[114] Mataveli LRV, Fioramonte M, Gozzo FC, Arruda MAZ. Improving Metallomics Information Related to Transgenic and Non-Transgenic Soybean Seeds Using 2D-HPLC-ICP-MS and ESI-MS/MS. Metallomics 2012;4(4) 373-378.

[115] Cozzolino R, de Giulio B. Application of ESI and MALDI-TOF-MS for Triacylglycerols Analysis in Edible Oils. European Journal of Lipid Science and Technology 2011;113(2) 160-167.

[116] Pessôa GS, Pilau EJ, Gozzo FC, Arruda MAZ. Ion Mobility Mass Spectrometry: an Elegant Alternative Focusing on Speciation Studies. Journal of Analytical and Atomic Spectrometry 2011;26(1) 201-206.

[117] Elliot MH, Smith DS, Parkera CE, Borchers, C. Current Trends in Quantitative Proteomics. Journal of Mass Spectrometry 2009;44(12), 1637-1660.

[118] Rodríguez-Cea A, de la Campa MRF, Alonso, IG, Sanz-Medel A. The Use of Enriched 111Cd as Tracer to Study De Novo Cadmium Accumulation and Quantitative Speciation in Anguilla Anguilla Tissues. Journal of Analytical and Atomic Spectrometry 2006;21(3) 270-278.

[119] Vanhaecke F., Balcaen L., Taylor P. Use of ICP-MS for Isotope Ratio Measurements. In: Hill S.J. (ed.) Inductively Coupled Plasma Spectrometry and its Applications. Oxford: Blackwell Publishing; 2007. p160-225.

[120] Vanhaecke F, Balcaen L, Malinovsky, D. Use of Single-Collector and Multi-Collector ICP-Mass Spectrometry for Isotopic Analysis. Journal of Analytical and Atomic Spectrometry 2009;24(7) 863-886.

[121] Wu B, Zoriy M, Chen Y, Becker JS. Imaging of Nutrient Elements in the Leaves of Elsholtzia Splendens by Laser Ablation Inductively Coupled Plasma Mass Spectrometry (LA-ICP-MS). Talanta 2009;78(1) 132-137.

[122] Vanhaecke F., Kyser K. The Isotopic Composition of the Elements. In: Vanhaecke F., Degryse, P. (ed.) Isotopic Analysis: Fundamentals and Application Using ICP-MS. Weinheim: Wiley VCH; 2012. p1-29.

Permissions

The contributors of this book come from diverse backgrounds, making this book a truly international effort. This book will bring forth new frontiers with its revolutionizing research information and detailed analysis of the nascent developments around the world.

We would like to thank James E. Board, for lending his expertise to make the book truly unique. He has played a crucial role in the development of this book. Without his invaluable contribution this book wouldn't have been possible. He has made vital efforts to compile up to date information on the varied aspects of this subject to make this book a valuable addition to the collection of many professionals and students.

This book was conceptualized with the vision of imparting up-to-date information and advanced data in this field. To ensure the same, a matchless editorial board was set up. Every individual on the board went through rigorous rounds of assessment to prove their worth. After which they invested a large part of their time researching and compiling the most relevant data for our readers. Conferences and sessions were held from time to time between the editorial board and the contributing authors to present the data in the most comprehensible form. The editorial team has worked tirelessly to provide valuable and valid information to help people across the globe.

Every chapter published in this book has been scrutinized by our experts. Their significance has been extensively debated. The topics covered herein carry significant findings which will fuel the growth of the discipline. They may even be implemented as practical applications or may be referred to as a beginning point for another development. Chapters in this book were first published by InTech; hereby published with permission under the Creative Commons Attribution License or equivalent.

The editorial board has been involved in producing this book since its inception. They have spent rigorous hours researching and exploring the diverse topics which have resulted in the successful publishing of this book. They have passed on their knowledge of decades through this book. To expedite this challenging task, the publisher supported the team at every step. A small team of assistant editors was also appointed to further simplify the editing procedure and attain best results for the readers.

Our editorial team has been hand-picked from every corner of the world. Their multi-ethnicity adds dynamic inputs to the discussions which result in innovative outcomes. These outcomes are then further discussed with the researchers and contributors who give their valuable feedback and opinion regarding the same. The feedback is then collaborated with the researches and they are edited in a comprehensive manner to aid the understanding of the subject.

Apart from the editorial board, the designing team has also invested a significant amount of their time in understanding the subject and creating the most relevant covers. They scrutinized every image to scout for the most suitable representation of the subject and create an appropriate cover for the book.

The publishing team has been involved in this book since its early stages. They were actively engaged in every process, be it collecting the data, connecting with the contributors or procuring relevant information. The team has been an ardent support to the editorial, designing and production team. Their endless efforts to recruit the best for this project, has resulted in the accomplishment of this book. They are a veteran in the field of academics and their pool of knowledge is as vast as their experience in printing. Their expertise and guidance has proved useful at every step. Their uncompromising quality standards have made this book an exceptional effort. Their encouragement from time to time has been an inspiration for everyone.

The publisher and the editorial board hope that this book will prove to be a valuable piece of knowledge for researchers, students, practitioners and scholars across the globe.

List of Contributors

Thankaraj Salammal Mariashib, Vasudevan Ramesh Anbazhagan, Shu-Ye Jiang and Srinivasan Ramachandran
Temasek Life Sciences Laboratory, 1 Research Link, the National University of Singapore, Singapore

Andy Ganapathi
Department of Biotechnology, Bharathidasan University, Tiruchirappalli, Tamilnadu, India

Alka Dwevedi
Regional Centre for Biotechnology, India

Arvind M Kayastha
School of Biotechnology, Faculty of Science, Banaras Hindu University, India

Laura C. Hudson, Kevin C. Lambirth, Kenneth L. Bost and Kenneth J. Piller
University of North Carolina at Charlotte and SoyMeds, Inc., USA

Murilo Siqueira Alves and Luciano Gomes Fietto
Department of Biochemistry and Molecular Biology, Federal University of Viçosa, Viçosa, Minas Gerais, Brazil

Megumi Kasai, Mayumi Tsuchiya and Akira Kanazawa
Research Faculty of Agriculture, Hokkaido University, Japan

Cristiane Fortes Gris
Federal Institute of Southern Mines, Brazil

Alexana Baldoni
Federal University of Lavras, Brazil

Lidia Skuza, Ewa Filip and Izabela Szućko
Cell Biology Department, Faculty of Biology, University of Szczecin, Poland

Hyeyoung Lee, So-Yon Park and Zhanyuan J. Zhang
Plant Transformation Core Facility, University of Missouri, Columbia, USA

Marco Aurélio Zezzi Arruda, Herbert de Sousa Barbosa, Lidiane Raquel Verola Mataveli and Silvana Ruella Oliveira
Institute of Chemistry, National Institute of Science and Technology for Bio analytics and Department of Analytical Chemistry, University of Campinas, Campinas, Brazil

Ricardo Antunes Azevedo and Sandra Cristina Capaldi Arruda
Department of Genetics, Laboratory of Genetics Biochemistry of Plants, University of São Paulo, ESALQ, Piracicaba, Brazil

Priscila Lupino Gratão
Department of Applied Biology to Agricultural, Universidade Estadual Paulista, FCAV, Jaboticabal, Brazil